ESSAYS IN BIOCHEMISTRY

Other recent titles in the *Essays in Biochemistry* series:
The Role of Non-Coding RNAs in Biology: volume 54
edited by M.A. Lindsay and S. Griffiths-Jones
2013
ISBN 978 1 85578 190 0

Cell Polarity and Cancer: volume 53
edited by A.D. Chalmers and P. Whitley
2012
ISBN 978 1 85578 189 4

Lysine-Based Post-Translational Modification of Proteins: volume 52
edited by I. Scott
2012
ISBN 978 1 85578 185 6

Molecular Parasitology: volume 51
edited by R. Docampo
2011
ISBN 978 1 85578 184 9

ABC Transporters: volume 50
edited by F.J. Sharom
2011
ISBN 978 1 85578 181 8

Chronobiology: volume 49
edited by H.D. Piggins and C. Guilding
2011
ISBN 978 1 85578 180 1

Epigenetics, Disease and Behaviour: volume 48
edited by H.J. Lipps, J. Postberg and D.A. Jackson
2010
ISBN 978 1 85578 179 5

ESSAYS IN BIOCHEMISTRY

volume 55 2013

Autophagy: Molecules and Mechanisms

Edited by Jon D. Lane

Series Editor
Nigel Hooper (Leeds, U.K.)

Advisory Board
G. Banting (Bristol, U.K.)
E. Blair (Leeds, U.K.)
P. Brookes (Rochester, NY, U.S.A.)
S. Gutteridge (Newark, DE, U.S.A.)
J. Pearson (London, U. K.)
J. Rossjohn (Melbourne, Australia)
E. Shephard (London, U.K.)
J. Tavaré (Bristol, U.K.)
C. Tournier (Manchester, U.K.)

Essays in Biochemistry is published by Portland Press Limited on behalf of the Biochemical Society

Portland Press Limited
Third Floor, Charles Darwin House
12 Roger Street
London WC1N 2JU
U.K.
Tel: +44 (0)20 7685 2410
Fax: +44 (0)20 7685 2469
email: editorial@portlandpress.com
www.portlandpress.com

© The Authors; Journal compilation © 2013 Biochemical Society

All rights reserved. Apart from any fair dealing for the purposes of research or private study, or criticism or review, as permitted under the Copyright, Designs and Patents Act, 1998, this publication may be reproduced, stored or transmitted, in any forms or by any means, only with the prior permission of the publishers, or in the case of reprographic reproduction in accordance with the terms of the licences issued by the Copyright Licensing Agency. Inquiries concerning reproduction outside those terms should be sent to the publishers at the above-mentioned address.

Although, at the time of going to press, the information contained in this publication is believed to be correct, neither the authors nor the editors nor the publisher assumes any responsibility for any errors or omissions herein contained. Opinions expressed in this book are those of the authors and are not necessarily held by the Biochemical Society, the editors or the publisher.

All profits made from the sale of this publication are returned to the Biochemical Society for the promotion of the molecular life sciences.

British Library Cataloguing-in-Publication Data
A catalogue record for this book is available from the British Library
ISBN   978-1-85578-191-7
ISSN (print)   0071 1365
ISSN (online)   1744 1358

Typeset by Techset Composition Ltd, Salisbury, U.K.
Printed in Great Britain by Cambrian Printers Ltd, Aberystwyth

CONTENTS

Preface xi

Authors xv

Abbreviations xix

1 Early signalling events of autophagy 1
Laura E. Gallagher and Edmond Y.W. Chan

Abstract 1
Introduction 1
Autophagosome initiation, elongation and closure 3
Regulation of yeast autophagy by TOR–ATG1 3
ATG1 functional domains 6
Regulation of the mammalian ULK1 complex 7
Beyond TOR–ATG1 8
AMPK regulation of autophagy 8
Multiple mTORC1–AMPK–ULK1 connections 9
Post-translational and transcriptional control of ULK1 9
Downstream of ATG1/ULK1 10
ATG1/ULK1 for neurobiology 11
Conclusion 11
Summary 12
References 12

2 Omegasomes: PI3P platforms that manufacture autophagosomes 17
Rebecca Roberts and Nicholas T. Ktistakis

Abstract 17
Introduction 17
Autophagy 18
Omegasomes 20
Regulation of omegasome formation 22
Omegasomes during pathogen-induced autophagy 23
Termination of the PI3P signal: role for several 3-phosphatases 24
Conclusion 25
Summary 25
References 25

© 2013 Biochemical Society

3 Current views on the source of the autophagosome membrane 29
Sharon A. Tooze

Abstract 29
Introduction 30
Definition of the PAS and phagophore 31
The molecular machinery 31
The function of Atg proteins: hierarchal analysis 33
The origin and source of the phagophore 34
Conclusion 36
Summary 36
References 36

4 Two ubiquitin-like conjugation systems that mediate membrane formation during autophagy 39
Hitoshi Nakatogawa

Abstract 39
Introduction 40
Conjugation reactions of Atg12 and Atg8 41
Functions of Atg12–Atg5 and Atg8–PE conjugates 42
Mechanism of Atg3 activation by Atg12–Atg5 44
Spatial regulation of Atg8–PE formation by the
   Atg12–Atg5–Atg16 complex 45
Significance of Atg8 deconjugation by Atg4 46
Conclusions 46
Summary 47
References 48

5 The Atg8 family: multifunctional ubiquitin-like key regulators of autophagy 51
Moran Rawet Slobodkin and Zvulun Elazar

Abstract 51
Introduction 52
The Atg8 family 53
Atg8 processing 54
Function of the Atg8 family 56
LC3 as a tool to monitor the autophagic process 59
Conclusions and future aspects 60
Summary 61
References 62

6 Autophagosome maturation and lysosomal fusion 65
Ian G. Ganley

Abstract 65
Introduction 65
Maturation of the autophagosome 69
Travelling to endosomes and lysosomes: role of the cytoskeleton 69
Fusion of the autophagosome 71
Rabs: master coordinators of membrane trafficking 72
Membrane tethers: bridges to fusion 74
SNAREs: the driving force of membrane fusion 74
Conclusion 75
Summary 75
References 76

7 Selective autophagy 79
Steingrim Svenning and Terje Johansen

Abstract 79
Introduction 79
Autophagy receptors 81
Interaction of autophagy receptors with ATG8/LC3 84
Ubiquitin-mediated degradation 85
Selective autophagy of protein aggregates: aggrephagy 86
Xenophagy 88
Regulation of selective autophagy 88
Concluding remarks 89
Summary 89
References 90

8 Mitophagy 93
Thomas MacVicar

Abstract 93
Introduction 93
Mitophagy mechanisms 94
Mitophagy in yeast 94
Mitophagy during erythropoiesis 95
Parkin-mediated mitophagy and beyond 96
Mitophagy regulation by mitochondrial dynamics 99
Mitochondrial dynamics: a balance between fusion and fission 99
Linking fission with mitophagy 99
Fusion protects mitochondria from mitophagy 100
Conclusion 100
Summary 101
References 101

© 2013 Biochemical Society

9 Autophagy and cell death 105
Tohru Yonekawa and Andrew Thorburn

Abstract 105
Introduction 105
Autophagy as a promoter of cell death 107
Autophagy as a protector against cell death 109
Concluding remarks 113
Summary 114
References 114

10 Autophagy and ageing: implications for age-related neurodegenerative diseases 119
Bernadette Carroll, Graeme Hewitt and Viktor I. Korolchuk

Abstract 119
Introduction 120
Observed changes in autophagy during ageing 122
The contribution of autophagy to ageing 125
Autophagy in age-related neurodegeneration 126
Conclusion 128
Summary 129
References 129

11 Role of autophagy in cancer prevention, development and therapy 133
G. Vignir Helgason, Tessa L. Holyoake and Kevin M. Ryan

Abstract 133
Introduction 134
Molecular regulators of mammalian autophagy 135
Signalling regulation of mammalian autophagy 136
Growth factor signalling: RAS and the PI3K–Akt–mTORC1 pathway 136
Energy sensing: positive regulation by AMPK 138
Stress response: the dual role of p53 138
The link to cell death: Bcl-2 protein family 138
The paradoxical role of autophagy in cancer 139
Autophagy therapeutics 144
Autophagy induction in cancer prevention and improved chemotherapy 144
Autophagy inhibition in combination with anticancer therapy 145
Future directions 146
Conclusions 146
Summary 147
References 147

12 Autophagy as a defence against intracellular pathogens 153
Tom Wileman

Abstract 153
Introduction 154
The process of autophagy 154
Microbes that evade lysosomes are captured by autophagy 155
Autophagy is activated following recognition of pathogen-associated molecular patterns and damage signals 157
Selective autophagy involves autophagy receptors with LC3-interacting regions 158
Autophagy receptors control selective autophagy of intracellular pathogens 158
Microbial evasion of autophagy 160
Conclusions and future research 161
Summary 161
References 162

Index 165

PREFACE

Autophagy has emerged in recent years as one of the most exciting fields in cell biology. From its meagre beginnings as a curiosity reported by the pioneers of electron microscopy in the 1950s, through its first description in 1963 by the late Nobel Laureate, Christian DeDuve, to the latter day application of yeast genetics, cell biology and vertebrate model organism research, interest in this key cellular process has blossomed. Until quite recently, the mechanistic control of autophagy remained obscure, but with the molecular jigsaw pieces now being assembled at a terrific pace, there is a need to take stock and consider how far we have come and how much there is still to learn. This is particularly important for autophagy because very little background information is available in textbooks for consultation by students and early researchers interested in learning about this field.

To chart the increasing numbers of articles on autophagy in major journals over recent years tells only part of the story of the expanding interest in this burgeoning field. Only when one delves into the diversity of the scientific disciplines for which autophagy contributes, can one fully appreciate the broad influence of this essential biological process. Autophagy has roles in development, tissue and cell differentiation and maturation, cell division, homoeostasis and metabolic control, to name but a few. Appropriately enough, dysregulation of autophagy is now known to cause or contribute to a range of human diseases, further underlining the importance of studying the regulation of this pathway, and further explaining the explosion in interest in this topic.

Three major types of autophagy have been characterized to date, CMA (chaperone-mediated autophagy), microautophagy and macroautophagy. This collection of essays is dedicated to macroautophagy, the most studied and thus best understood form of autophagy, and the reader is directed to other sources for coverage of the other autophagy pathways. In this preface, and in the following chapters, the common practice of describing macroautophagy as simply 'autophagy' will be adopted. Autophagy is, strictly speaking, a membrane trafficking process. It describes the *de novo* assembly and maturation of novel double-membrane organelles that expand and close to sequester cytoplasm, including whole organelles, in order to efficiently deliver surplus or redundant cellular cargo to the lysosomes for degradation and recycling. Driving this complex process of membrane assembly/remodelling, trafficking and heterotypic organelle fusion is a set of essential autophagy-specific proteins. These include protein and lipid kinases, membrane targeting and trafficking factors, endopeptidases, and protein and lipid ubiquitin-like modifiers. In this volume of *Essays in Biochemistry*, our current understanding of the events that control autophagosome assembly and maturation will be described (Chapters 1–7), before the influence of autophagy during ageing, infection and disease is examined (Chapters 9–12). Separating these, are two chapters (Chapters 7 and 8) that discuss how autophagy can become highly selective to degrade specific cytoplasmic cargoes, a key facet of cellular protein and organelle quality control.

© 2013 Biochemical Society

In Chapter 1, "Early signalling events of autophagy", Laura Gallagher and Ed Chan explain how the autophagy-specific protein kinase, ATG1 (or ULK1 in mammalian cells) responds to changing nutrient levels via mTOR (mammalian/mechanistic target of rapamycin) signalling to trigger the cascade of downstream events that drive autophagosome biogenesis. Chapter 2 by Rebecca Roberts and Nicholas Ktistakis, "Omegasomes: PI3P platforms that manufacture autophagosomes", focuses on a key step during autophagosome assembly, the activation of a class III phosphoinositide 3-kinase that generates phosphatidylinositol 3-phosphate (PI3P) at the ER (endoplasmic reticulum) to enable recruitment of downstream effectors. These PI3P-rich membranes sub-domains expand to adopt an idiosyncratic Ω shape, and are thus described as 'omegasomes'. In Chapter 3, Sharon Tooze reviews current theories on the origins of membranes for autophagosome assembly, still a controversial area in the autophagy field. Chapters 4 and 5 focus on the mechanistic pathways that control the assembly of the autophagosome. In Chapter 4, "Two ubiquitin-like conjugation systems that mediate membrane formation during autophagy", Hitoshi Nakatogawa explains how conjugation of ATG12 to ATG5 and LC3/ATG8 to the lipid, phosphatidylethanolamine, controls the expansion and fusion of the nascent autophagosome. In Chapter 5, "The ATG8 family: multifunctional ubiquitin-like key regulators of autophagy", Moran Slobodkin and Zvulun Elazar focus on the only stably integrated protein constituents of the autophagosomal limiting membrane, the ATG8 family of ubiquitin-like lipid modifiers. They discuss how cleavage of ATG8 family members at conserved C-terminal sites by the ATG4 family of endopeptidases is crucial for ATG8 function. To understand the physiological significance of autophagy, and to appreciate how failures in autophagy can contribute to disease, it is essential that the efficiency of the lysosomal fusion step be considered. In Chapter 6, "Autophagosome maturation and lysosomal fusion", Ian Ganley examines the events that take place during autophagosome maturation, how autophagosomes merge with the endocytic compartment en route to the lysosome, and the roles played by the cytoskeleton, Rabs, and membrane tethers and SNAREs.

The next two chapters are dedicated to selective autophagy. In Chapter 7, "Selective autophagy", Steingrim Svenning and Terje Johansen discuss how the regulated binding of adaptor molecules to ATG8 family members allows the efficient incorporation of diverse cellular cargoes into the nascent autophagosome. They describe how this process enables cells to degrade protein aggregates, redundant/damaged organelles and invading microorganisms, a topic that is revisited in the final chapter. In Chapter 8, Thomas MacVicar describes the varied pathways that allow the selective degradation of mitochondria by autophagy. This process is described as "Mitophagy", and is essential for mitochondrial quality control and for mitochondrial removal during erythroid differentiation.

The final four chapters in the book examine how autophagy influences cell viability in health and disease. In Chapter 9, "Autophagy and cell death", Tohru Yonekawa and Andrew Thorburn discuss the ways in which autophagy can both contribute to and protect against cell death in a tissue- and context-specific manner.

© The Authors Journal compilation © 2013 Biochemical Society

In doing so, they describe the key proteins that provide a mechanistic link between autophagy and apoptosis. In Chapter 10, "Autophagy and ageing: implications for age-related neurodegenerative diseases", Bernadette Carroll, Graeme Hewitt and Viktor Korolchuk describe the roles that autophagy play during the ageing process, focusing on how autophagy pathways (including CMA) prevent the accumulation of toxic protein aggregates that are known to contribute to neuronal cell death during neurodegenerative disease. In Chapter 11, "Role of autophagy in cancer prevention, development and therapy", G. Vignir Helgason, Tessa Holyoake and Kevin Ryan describe how the signalling pathways that are disrupted in cancer influence the autophagy pathway, and how autophagy can be both oncogenic and tumour suppressive, depending on context. The authors then go on to examine how autophagy might be targeted for cancer therapy. In Chapter 12, "Autophagy as a defence against intracellular pathogens", Tom Wileman describes the essential role played by autophagy in recognizing and degrading invading microorganisms, an important facet of host defences against pathogens.

Finally, I would like to acknowledge the people who have made this book possible. I am indebted to the Portland Press staff, in particular, Clare Curtis who has coordinated this process in a highly efficient and professional manner. I am grateful to the anonymous reviewers of the initial proposal and of the chapters themselves, and of course I thank the authors for writing such stimulating and high-quality reviews.

AUTHORS

Jon Lane is a Reader in Cell Biology at the University of Bristol, U.K. He obtained a Ph.D. at the University of Exeter in the mid-1990s, using insect and frog model systems to study the regulation of microtubule organization and stability through the cell cycle and during development. He then did postdoctoral research in Manchester, studying microtubule motor proteins and microtubule-based membrane movement using frog egg and embryo extracts. During this time, he became interested in cell death (apoptosis), and how dying cells re-organize their cytoskeletal and organelle systems to facilitate recognition and removal by cells of the innate immune system. Moving to Bristol in 2003 as a Wellcome Trust Career Development Fellow, then later as a Research Councils UK Fellow, he developed a strong interest in autophagy. His group is now focused on autophagy, with the aim of understanding how autophagy regulates cellular function during development and disease.

Laura Gallagher graduated with a first class honours Master of Pharmacology degree from the University of Bath in 2012. She is currently a Ph.D. student working with Dr Edmond Chan characterizing the role of autophagy during breast cancer metastasis.

Edmond Chan is a Lecturer of Biochemistry in the Faculty of Science, University of Strathclyde. Dr Chan has research interests in the signalling mechanisms that coordinate the multiple stages of autophagy. From 2003 to 2008, Dr Chan worked at the Cancer Research UK London Research Institute, where he screened siRNA libraries for autophagy factors and began focusing on nutrient-dependent kinases. Before that, he studied neurodegeneration using mouse models of Huntington's disease at the University of British Columbia, Canada. Dr Chan obtained his Ph.D. in Biochemistry (1999) from the University of Alberta, Canada.

Rebecca Roberts is a Research Associate in the Ktistakis laboratory in the signalling department at the Babraham Research Institute, Cambridge. For her Ph.D. from the University of East Anglia in Norwich, Rebecca studied the activation of autophagy with viral and non-viral particles. In her current research, Rebecca is searching for novel regulators of autophagy using an siRNA genome library.

Nicholas Ktistakis has been a group leader at the Signalling Programme of the Babraham Research Institute since 1996. He studies lipid signalling, with special emphasis on pathways regulated by phosphatidic acid and phosphatidylinositol 3-phosphate.

Sharon A. Tooze is a senior scientist at the London Research Institute, Cancer Research UK. Dr Tooze completed her Ph.D. and postdoctoral training at the European Molecular Biology Laboratory in Heidelberg, Germany, studying membrane trafficking in the Cell Biology Programme. Dr Tooze established her own laboratory at the London Research Institute in 1993 to pursue her interest in organelle biogenesis, first addressing secretory granule biogenesis. She began studying autophagy in 2006.

Hitoshi Nakatogawa received his Ph.D. from Kyoto University, Kyoto, Japan, in 2002, where he studied the regulation of protein synthesis and secretion in bacteria. Then, he joined Yoshinori Ohsumi's group at the National Institute for Basic Biology, Okazaki, as a postdoc in 2004, and the group moved to the Tokyo Institute of Technology, Yokohama, in 2009. He is currently an associate professor in the same group and is working on the molecular mechanisms of autophagy in yeast with a special focus on ubiquitin-like conjugation systems.

Moran Rawet Slobodkin is currently a postdoctoral fellow in the Elazar laboratory at the Weizmann Institute of Science. She completed her Ph.D. studies at the Faculty of Biology, Technion, Haifa in Professor Cassel's laboratory where she studied the mechanism of action of ARF-GAPs in membrane trafficking. Her present research focuses on the regulation of autophagy under different growth conditions.

Zvulun Elazar obtained his Ph.D. at the Weizmann Institute of Science in Rehovot, Israel. Currently, he is the Harold Korda Professor at the Department of Biological Chemistry of the Weizmann Institute of Science where he studies different mechanistic aspects of autophagy in yeast and mammalian systems. He is particularly interested in the role of the Atg8s, unique ubiquitin-like proteins that are part of the core autophagic machinery. The current focus of his laboratory is on the identification of factors that regulate these processes and are defective in pathophysiological conditions such as neurodegeneration and cancer.

Ian Ganley obtained his Ph.D. from the University of Cambridge working on the enzyme phospholipase D with Dr Nicholas Ktistakis. This was followed up with postdoctoral work in the laboratories of Professor Suzanne Pfeffer and Dr Xuejun Jiang at Stanford University and Memorial Sloan-Kettering Cancer Center respectively. It was there that Ian developed a passion for intracellular transport and how this relates to autophagy. In 2010, Ian relocated to Dundee to start his own laboratory focusing on the mechanisms of autophagy regulation.

Steingrim Svenning is currently working on his Ph.D. thesis as a member of Terje Johansen's group. His main work lies in characterizing the role of the selective autophagy receptor protein NBR1.

Terje Johansen is a Professor and group leader at the Institute of Medical Biology at the University of Tromsø, Norway. During work on cell signalling on atypical protein kinase C observations on the behaviour of an interacting protein p62/SQSTM1 led to the discovery of p62/SQSTM1 as the first selective autophagy receptor. Since then his focus has been set on the autophagy receptor proteins, as well as other aspects of selective autophagy.

Thomas MacVicar graduated with a B.Sc. in Biochemistry from the University of Bristol in 2009. He is currently studying for a Ph.D. in Jon Lane's laboratory (School of Biochemistry, University of Bristol) and is conducting research on mitophagy regulatory mechanisms.

Tohru Yonekawa is a postdoctoral fellow at the University of Colorado. His current research focuses on understanding signalling pathways that differentially regulate autophagy.

Andrew Thorburn is Professor and Chair of the Pharmacology Department and Deputy Director of the Cancer Center at the University of Colorado, School of Medicine. His research focuses on the role of autophagy and its interplay with apoptosis in cancer progression and treatment.

Bernadette Carroll recently completed her Ph.D. at Imperial College London. She is now working as a postdoctoral associate at Newcastle University investigating cellular signalling in response to nutrients.

Graeme Hewitt completed his undergraduate and Master's studies in Biomedicine with a focus on ageing. He is currently studying for his Ph.D. at Newcastle University investigating DNA damage and its associations with autophagy and ageing.

Viktor Korolchuk obtained his Ph.D. in Kiev, Ukraine, followed by postdoctoral training at Bristol and Cambridge Universities. He is currently a Lecturer at Newcastle University, studying nutrient-dependent signalling and trafficking pathways and their relevance to ageing.

G. Vignir Helgason trained for a Ph.D. degree in Professor Ryan's laboratory at the Beatson Institute and graduated from the University of Glasgow in October 2007. Since 2007, Dr Helgason held postdoctoral positions in Professor Holyoake's laboratory at the Paul O'Gorman Leukaemia Research Centre and then became Kay Kendall Leukaemia Fund Intermediate Research Fellow in 2013. Dr Helgason has also held a Lord Kelvin Adam Smith Leadership Fellowship from University of Glasgow since 2013. His research currently focuses on the role of autophagy in disease persistence and drug resistance in chronic myeloid leukaemia.

Tessa Holyoake is Director of the Paul O'Gorman Leukaemia Research Centre, University of Glasgow and a Consultant Haematologist at the West of Scotland Cancer Centre. Her hypothesis, driven translational research on cancer stem cells in chronic myeloid leukaemia, is of international standing and she is widely recognized as one of the key players in this field. This research focus developed from her Ph.D. at the Beatson Institute, Glasgow, on haemopoietic stem cell expansion for therapeutic use, through a 2-year postdoctoral fellowship in the Terry Fox Laboratories, British Columbia, to the current cancer stem cell focus. Professor Holyoake was made a fellow of the Royal Society of Edinburgh in 2008, was awarded the Scottish Health Award in Cancer in 2009 and the Lord Provost's award for Health in 2010.

Kevin Ryan received his Ph.D. from the Beatson Institute, University of Glasgow in 1996. Following postdoctoral work at the US National Cancer Institute, he was awarded a Cancer Research UK Senior Cancer Research Fellowship in 2002 and returned to the Beatson Institute to establish his own research group. In 2007, he was promoted to Senior Group Leader at the Beatson Institute and later that year was appointed Professor of Molecular Cell Biology at the University of Glasgow. Kevin's studies focus on the identification of cell death regulators involved in tumour development and tumour therapy with particular focus on how the apoptotic and autophagic pathways integrate to determine cell fate. In recognition of his work, he was awarded the 2010 European Association for Cancer Research (EACR) 'Cancer Researcher Award' and the 2012 Tenovus Medal.

Tom Wileman is Professor of Molecular Virology at the Norwich Medical School at the University of East Anglia. He trained in cell biology and immunology at Washington University and Harvard Medical Schools in the U.S.A. between 1982 and 1994. During this period he studied receptor-mediated endocytosis and endoplasmic reticulum-related protein degradation. In 1994 he moved to the Institute for Animal Health Pirbright Laboratories in the U.K. and continued his interests in membrane traffic by studying the role played by membrane compartments in the control of virus infection and replication. His current interests at Norwich Medical School focus on understanding the role played by autophagy in controlling the replication of viruses, particularly how viral proteins activate autophagy, and the consequences this has for the survival of viruses within cells.

ABBREVIATIONS

AIM	ATG8 family-interacting motif
ALFY	autophagy-linked FYVE
AMBRA1	activating molecule in BECN1-regulated autophagy 1
AMPK	AMP-activated protein kinase
Asn*	invariable asparagine residue
ATF4	activating transcription factor 4
ATG/Atg	autophagy gene product/autophagy-related protein
BAD	Bcl-2/Bcl-x-L antagonist, causing cell death
BH	Bcl-2 homology
BIK	Bcl-2-interacting killer
BIM	Bcl-2-interacting mediator of cell death
CMA	chaperone-mediated autophagy
CML	chronic myeloid leukaemia
COPI	coatomer protein I
CQ	chloroquine
CR	caloric restriction
Cvt	cytoplasm-to-vacuole targeting
DAG	diacylglycerol
DAMP	damage-associated molecular pattern
dBRUCE	*Drosophila* baculovirus inhibitor of apoptosis repeat-containing ubiquitin-conjugating enzyme
DFCP1	double FYVE domain-containing protein 1
DISC	death-inducing signalling complex
DRAM-1	damage-regulated autophagy modulator 1
DUB	de-ubiquitinating enzyme
EAT	early autophagy targeting/tethering
EM	electron microscopy
ER	endoplasmic reticulum
ESCRT	endosomal sorting complex required for transport
FADD	Fas-associated death domain
FIP200	focal adhesion kinase family-interacting protein 200 kDa
FMDV	Foot and Mouth Disease virus
FYCO1	FYVE and coiled-coil domain containing 1
GABARAP	γ-aminobutyric acid receptor-associated protein
GAP	GTPase-activating protein
GATE-16	Golgi-associated ATPase enhancer of 16 kDa
GBM	glioblastoma multiforme
GEF	guanine nucleotide-exchange factor
gp78	glycoprotein 78
GSK3	gycogen synthase kinase 3

HCQ	hydroxychloroquine
HDAC6	histone deacetylase 6
HOPS	homotypic fusion and vacuolar protein sorting
HSC	haemopoietic stem cell
Hsc70	heat-shock cognate 70 stress protein
Hsp	heat-shock protein
IM	imatinib
IMM	inner mitochondrial membrane
KEAP1	Kelch-like ECH (erythroid cell-derived protein with cap 'n' collar homology)-associated protein 1
LC3	light-chain 3
LIR	LC3-interacting region
LLR	leucine rich
MAM	mitochondria-associated endoplasmic reticulum membrane
MCL-1	myeloid cell leukaemia sequence 1
MER	minimal essential region
Mfn	mitofusin
MTM	myotubularin
mTOR	mammalian/mechanistic target of rapamycin
mTORC	mammalian/mechanistic target of rapamycin complex
NBR1	neighbour of BRCA1 (breast cancer early-onset 1) gene 1
NDP52	nuclear dot protein 52
NIX	NIP3-like protein X
NOD	nucleotide-binding and oligomerization domain
NSF	N-ethylmaleimide-sensitive factor
OMM	outer mitochondrial membrane
OPA1	optic atrophy 1
PAMP	pathogen-associated molecular pattern
PAS	phagophore assembly site/pre-autophagosomal structure
PB1	Phox and Bem1p
PD	Parkinson's disease
PE	phosphatidylethanolamine
PI3K	phosphoinositide 3-kinase
PI3P	phosphatidylinositol 3-phosphate
PINK1	PTEN (phosphatase and tensin homologue deleted on chromosome 10)-induced putative kinase 1
PIP_3	phosphatidylinositol 3,4,5-trisphosphate
proAPI	precursor aminopeptidase-I
PUMA	p53 up-regulated modulator of apoptosis
raptor	regulatory associated protein of mTOR
Rheb	Ras homologue enriched in brain
RILP	Rab-interacting lysosomal protein
ROS	reactive oxygen species

SCV	*Salmonella*-containing vacuole
SINV	Sindbis virus
SLR	sequestosome 1-like receptor
SMURF1	SMAD-specific E3 ubiquitin ligase 1
SNAP-25	25 kDa synaptosome-associated protein
SNARE	soluble *N*-ethylmaleimide-sensitive factor-attachment protein receptor/soluble *N*-ethylmaleimide-sensitive fusion protein-attachment protein receptor
SQSTM1	sequestosome 1
Stbd1	starch-binding-domain-containing protein 1
TBK	TANK (tumour-necrosis-factor-receptor-associated factor-associated nuclear factor-κB activator)-binding kinase
TECPR	tectonin β-propeller repeat-containing protein 1
Tip60	HIV-1 Tat (transactivator of transcription)-interactive protein 60 kDa
TIRF	total internal reflection fluorescence
TLR	Toll-like receptor
TOR	target of rapamycin
TRAIL	tumour-necrosis-factor-related apoptosis-inducing ligand
TSC	tuberous sclerosis complex
UBA	ubiquitin-associated
Ubl	ubiquitin-like
ULK	uncoordinated-51-like kinase
UPS	ubiquitin–proteasome system
UVRAG	UV radiation resistance-associated gene
VMP1	vacuolar membrane protein 1
v-SNARE	vesicle-associated *N*-ethylmaleimidesensitive factor attachment protein receptor
WIPI	WD-repeat protein interacting with phosphoinositides

Early signalling events of autophagy

Laura E. Gallagher and Edmond Y.W. Chan[1]

Strathclyde Institute for Pharmacy and Biomedical Sciences, University of Strathclyde, Glasgow, Scotland G4 0RE, U.K.

Abstract

Autophagy is a conserved cellular degradative process important for cellular homoeostasis and survival. An early committal step during the initiation of autophagy requires the actions of a protein kinase called ATG1 (autophagy gene 1). In mammalian cells, ATG1 is represented by ULK1 (uncoordinated-51-like kinase 1), which relies on its essential regulatory cofactors mATG13, FIP200 (focal adhesion kinase family-interacting protein 200 kDa) and ATG101. Much evidence indicates that mTORC1 [mechanistic (also known as mammalian) target of rapamycin complex 1] signals downstream to the ULK1 complex to negatively regulate autophagy. In this chapter, we discuss our understanding on how the mTORC1–ULK1 signalling axis drives the initial steps of autophagy induction. We conclude with a summary of our growing appreciation of the additional cellular pathways that interconnect with the core mTORC1–ULK1 signalling module.

Keywords:
amino-acid starvation, AMPK, ATG1, ATG13, ATG17, FIP200, mTORC1, PAS, ULK.

Introduction

Autophagy is a conserved cellular process whereby damaged organelles and cytosolic proteins are degraded using double membrane-enclosed vesicles, known as autophagosomes [1]. The term 'autophagy' was coined in 1963 by Christian de Duve when he described the presence of lysosomal-like vesicles containing degraded cellular proteins and organelles. Since this

---

[1]To whom correspondence should be addressed (email Edmond.Chan@Strath.ac.uk).

discovery, three main types of autophagy have been described: chaperone-mediated autophagy, micro-autophagy and macro-autophagy. The most well understood of these is macro-autophagy, which will be the focus of this chapter and hereby simply referred to as autophagy. Autophagic degradation maintains cellular homoeostasis and promotes survival in almost every mammalian cell and this fundamental mechanism is integral to normal physiology and multiple disease states as summarized elsewhere [2–4]. In the present chapter, we discuss our understanding of the signalling events that are engaged when a cell activates autophagy.

The overall degradative flux of autophagy can be divided into three phases (Figure 1): first, sequestration, in which cellular components are captured by double-bilayer membranes to form the autophagosome; secondly, transport of the autophagosome to the lysosome; and thirdly, degradation or maturation, which involves vesicle fusion and mixing of autophagosomal contents with lysosomal hydrolases to eventually release degraded by-products back into the cytosol through membrane permeases. The production of autophagosomes is coordinated through the function of at least 35 ATG (autophagy) gene products, which were first

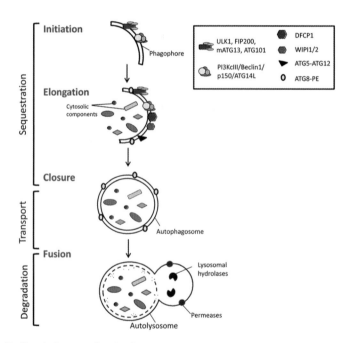

Figure 1. Defined phases of autophagy
Autophagy can be divided into three phases: sequestration of cellular components to form the autophagosome; transport of the autophagosome to the lysosome; and degradation. The initial process of sequestration is further divided into three stages, initiation, elongation and closure, which are dependent on sequential recruitment of ATG protein factors and membrane assembly. The ULK1 complex (ULK1, mATG13, FIP200 and ATG101) is involved at the earliest stage of autophagy initiation. The class III PI3K complex made up of Beclin1, PIKcIII (VPS34), ATG14L and p150 is similarly recruited to early forming autophagosomes. During elongation of the phagophore, further factors are recruited to the forming membrane, including PI3P effectors (WIPI1/2 and DFCP1), the ATG5–ATG12 conjugate and ATG8–PE, which all coordinate to promote closure of the complete autophagosome.

described in yeast, with following characterization of the mammalian orthologues (see [5] for detailed review).

Autophagosome initiation, elongation and closure

The initial sequestration step of autophagy can be differentiated further into three stages: initiation, elongation and closure. Initiation relies on the recruitment of multiple ATG protein complexes on to a cup-shaped membrane assembly site termed the phagophore (or isolation membrane). In mammalian cells, one of the earliest events during autophagosome initiation is the recruitment of the ULK1 (uncoordinated-51-like kinase 1) complex, which comprises ULK1, mATG13, FIP200 (focal adhesion kinase family-interacting protein 200 kDa) and ATG101 [6]. The homologous yeast complex, constructed around the kinase ATG1, is described in further detail below. A second complex recruited during autophagosome initiation contains the class III PI3K (phosphoinositide 3-kinase), VPS34, along with regulatory subunits Beclin1/ATG6, ATG14L and p150/VPS15. Activation of the Beclin1–PI3K complex results in production of PI3P (phosphatidylinositol 3-phosphate) to provide a lipid signal that recruits other autophagy effectors such as DFCP1 (double FYVE domain-containing protein 1) and members of the WIPI (WD-repeat protein interacting with phosphoinositides) family to the forming phagophore. The role of PI3K signalling for autophagy is discussed in a dedicated chapter of this volume (Chapter 2).

Initial phagophores elongate to enwrap cargo and this stage is dependent on recruitment and orchestrated action of the ATG5–ATG12 and ATG8–PE (phosphatidylethanolamine) ubiquitination-like conjugation systems. These two pathways are inter-related as the ATG12–ATG5 conjugate is required for progression of the second conjugation system, which ultimately converts ATG8 family proteins into their PE-conjugated (active) forms. Generation and enrichment of ATG8–PE on the phagophore has been proposed to drive a number of essential processes during autophagosome formation including binding of autophagy cargo receptors, membrane elongation and, finally, membrane closure.

ATG1, ATG6/Beclin1 and ATG8 represent the three core regulatory complexes essential for canonical autophagy, which describes the most widely observed forms of autophagy in mammalian cells [7]. The present chapter focuses on the regulation of canonical autophagy, which, characteristically, is rapidly activated in cells sensing starvation of nutrients such as amino acids (Figure 2). By contrast, other specialized non-canonical forms of autophagy can be activated in particular cell contexts. Non-canonical types of autophagy are notable for not requiring all of the major core regulatory modules, thus defining, for example, ATG1- or ATG8-independent subtypes of autophagy.

Regulation of yeast autophagy by TOR–ATG1

The best understood activation signal for autophagy is amino-acid starvation and this is transmitted using the TOR (target of rapamycin)–ATG1 pathway [8]. Cells use TOR–ATG1 signalling as a mechanism to coordinately regulate synthesis of proteins via translation and degradation via autophagy [9] (Figure 3). The key early studies showed how the

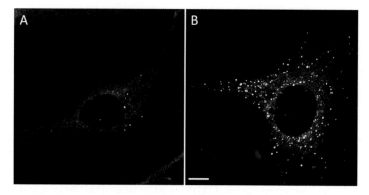

Figure 2. Nutrient-dependent autophagy in mammalian cells
Mouse embryonic fibroblasts were starved of amino acids for 2 h to stimulate the autophagy response. GFP fused to DFCP1 was used to mark early autophagosome assembly sites. Images show basal levels of autophagy (**A**) and induction following amino-acid starvation (**B**). Scale bar: 10 µm.

yeast *Saccharomyces cerevisiae* starved of amino acids increased ATG1 kinase activity to promote autophagy activation and this stimulatory effect was mimicked by the TOR inhibitor rapamycin [10]. Yeast lacking ATG1 could not undergo autophagy and could not endure starvation conditions showing the essential survival function for nutrient-dependent autophagy [11].

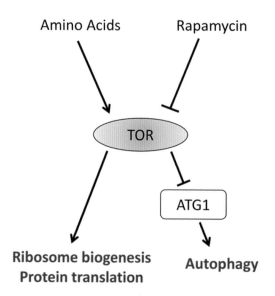

Figure 3. TOR balances autophagy and protein translation
Autophagy is controlled downstream of the central nutrient-sensitive kinase TOR, providing cells with a mechanism to balance protein synthesis and autophagy induction relative to amino acid availability. Under high amino acid concentrations, TOR inhibits ATG1 function, thus repressing autophagy, while promoting protein synthesis and cell growth. Amino-acid starvation or treatment with the drug rapamycin results in TOR inhibition, which allows ATG1 to become fully activated and signal autophagy induction.

© The Authors Journal compilation © 2013 Biochemical Society

Maximal activity of yeast ATG1 requires interaction with its regulatory cofactors ATG13, ATG17, ATG29 and ATG31 (Figure 4). In the current model, following autophagy activation, ATG1 and ATG13 assemble on to a stable ATG17–ATG29–ATG31 subcomplex [12]. Another key feature is that the function of ATG13 is controlled by phosphorylation. Under amino-acid-rich conditions, ATG13 is phosphorylated by TOR on at least eight serine residues [13]. Following amino-acid starvation and loss of TOR activity, ATG13 becomes hypo-phosphorylated, allowing full activation of ATG1 by stabilizing the overall complex or altering the conformation of ATG1–ATG13 binding. The importance of ATG1–ATG13–ATG17 complex formation has been demonstrated by mutagenesis approaches. The ATG17-C24R point mutation disrupts binding to ATG13 and this mutant was inefficient at supporting ATG1 kinase activity and autophagy in an ATG17 deletion strain [14]. Point mutations in the C-terminal region of ATG1 (Y878A/R885A and R885E/K892E) that prevent binding to

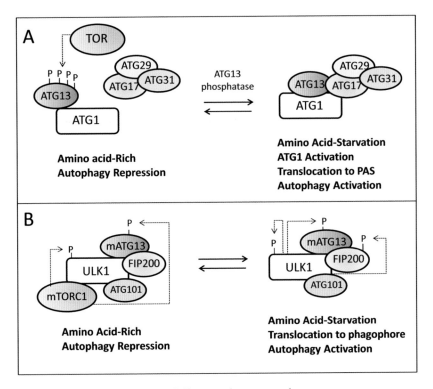

Figure 4. Assembly of ATG1 and ULK1 autophagy complexes
(**A**) Yeast: under amino-acid-rich conditions, ATG13 is phosphorylated by TOR on at least eight serine residues (a fraction of which are shown). Following amino-acid starvation and loss of TOR activity, ATG13 becomes hypo-phosphorylated, allowing full activation of ATG1 by potentially stabilizing the complex or altering the conformation of ATG1–ATG13 binding. In turn, ATG1 and ATG13 assemble on to a stable ATG17–ATG29–ATG31 subcomplex. The active ATG1 complex is assembled on the PAS to promote autophagosome formation. (**B**) Mammals: unlike yeast, the core mammalian ULK1 complex, comprising ULK1, mAtg13, ATG101 and FIP200, is stable independent of cellular amino acid levels. Under amino-acid-rich conditions, mTORC1 directly binds the ULK1 complex and phosphorylates mATG13 and ULK1, thereby inhibiting autophagy. During starvation, mTORC1 dissociates, allowing the active ULK1 complex to translocate on to autophagosome initiation sites.

© 2013 Biochemical Society

ATG13 accordingly prevent activation of autophagy [15]. Mutations in ATG13 that disrupt binding are similarly defective for autophagy regulation [16].

ATG1 functional domains

The initial membrane assembly site in yeast is specifically termed the PAS (pre-autophagosome structure). TOR inhibition following amino-acid starvation stimulates autophagy by promoting assembly of an active ATG1–ATG13–ATG17–ATG29–ATG31 complex on the PAS. Once localized on the PAS, multiple functional domains of ATG1 control separate immediate-early and late steps for autophagosome formation and maturation (Figure 5). Crystal structures of the ATG17–ATG29–ATG31 subcomplex have shown that dimerization of ATG17 may play a key role in bridging and coordinating the traffic of multiple lipid vesicle precursors at the PAS [17]. Part of this mechanism involves interaction with ATG1, which has as one of its key roles, direct binding of membrane vesicles via a C-terminal EAT (early autophagy targeting/tethering) domain. Within the C-terminal region of ATG1, the EAT domain appears overlapping or juxtaposed to the sequence that binds ATG13, leading speculatively to potential modes of co-operation or competition [15,18].

The ATG1 complex at the PAS is the pivotal director that recruits further autophagy factors. Consistent with this role, a defined sequence within the internal regulatory region of ATG1 binds ATG8, which has been termed the AIM (ATG8 family-interacting motif) [16,19]. Interestingly, mutation of the AIM did not prevent ATG1 localization and initial formation of the PAS, although ATG1–ATG8 interaction was critical for complete autophagosome formation and proper delivery to the yeast lysosomal-like vacuole compartment, highlighting multiple steps controlled by ATG1.

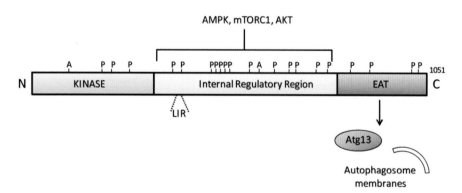

Figure 5. ATG1/ULK1 functional domains
Multiple functional domains of ATG1/ULK1 control separate immediate-early and late steps for autophagosome formation. The mammalian homologue ULK1 is shown as a representative of the ATG1 family. ULK1 contains an N-terminal kinase domain, an internal regulatory region and a C-terminal conserved region that contains the EAT domain. The EAT domain directs binding to membranes and is situated near or intermingled with sequences that bind the mATG13 cofactor. The internal regulatory domain contains a number of residues that are phosphorylated by kinases such as AMPK, mTORC1 and AKT. Within the internal regulatory domain lies the AIM (termed LIR in mammalian cells). ULK1 can be modified by other phosphorylation and acetylation (A) events.

In the current model, the fully assembled ATG1 complex is a fundamental component that provides a structural scaffold for early autophagosome formation. Other mutagenesis data further illustrate distinct ATG1 kinase-dependent and -independent roles during autophagosome formation [15]. Kinase-inactive ATG1 provided some function as the PAS was formed, but full progression of autophagy was impaired, for example, since factors like ATG8 abnormally accumulated. Consistent with a block in membrane flow, kinase-inactive ATG1 did not drive the generation of properly sized autophagosomes and functional autophagy. In summary, the first essential function of the ATG1 autophagy complex is to direct membranes and protein factors on to autophagosome assembly sites. A subsequent phase that requires ATG1 kinase function and ATG8 involves disassembly and remodelling of membrane and potentially other protein components to allow progression on to elongation, closure and transport phases of autophagy.

Regulation of the mammalian ULK1 complex

Mammalian genomes contain potentially five ATG1 orthologues, but of these, only ULK1 and ULK2 show conservation along their entire length [8]. To date, ULK1 is the best-characterized mammalian ATG1. As in yeast, the ULK1 complex transmits signals from the amino-acid-dependent mTORC1 (mechanistic TOR complex 1) to regulate autophagy. To avoid confusion, it is worth noting that mTOR was originally termed mammalian TOR, but has since been renamed mechanistic TOR, which is the terminology now recognized by the HUGO Gene Nomenclature Committee.

The internal regulatory region of ULK1 contains a conserved motif [microtubule-associated protein LIR (LC3-interacting region)] that binds to ATG8 family proteins [16,20] (Figure 5). The ULK1 C-terminus contains EAT- and mATG13-binding domains [18]. Interestingly, the mammalian mTORC1–ULK1 signalling module acquired additional novel mechanisms through evolution. In contrast with yeast, the core mammalian ULK1 complex (with mATG13, FIP200 and ATG101) remains stable regardless of the cellular nutrient status. Although there is no sequence similarity, FIP200 has been proposed to be the functional equivalent of ATG17 in the complex, whereas a yeast equivalent of ATG101 has yet to be characterized.

A critical feature of the mammalian signalling model is direct binding between activated mTORC1 and the ULK1 complex that is stimulated under amino-acid-rich conditions [6,8] (Figure 4). mTORC1 inhibits autophagy by phosphorylating ULK1 on Ser^{757} within the internal regulatory region and mTORC1-mediated phosphorylation of mATG13 could be providing a further inhibitory brake on autophagy [21–25]. Following amino-acid starvation, mTORC1 is inactivated and binding to the ULK1 complex is lost, thereby releasing the ULK1 complex to become catalytically active and translocation on to early autophagosome assembly sites near ER (endoplasmic reticulum) membranes. As in yeast, translocation of the ULK1 complex to phagophore initiation sites is an early event following amino-acid starvation in mammalian cells [25–27]. In the current model, ULK1 drives phagophore assembly and elongation by directing protein and membrane components. Without proper ULK function, other early steps of autophagosome assembly such as PI3P production from the Beclin1 complex are blocked [28]. ULK1 may play both structural and dynamic roles as kinase-inactive forms can be observed to localize to phagophore sites, but block overall autophagy flux [18,27]. Overall, the conserved TOR–ATG1 signalling module generally involves translocation of a multimeric ATG1 complex on to autophagosome assembly sites to orchestrate downstream protein and membrane traffic following nutrient starvation.

Beyond TOR–ATG1

The autophagy field is now in the process of further understanding how the basic TOR–ATG1 signalling axis is integrated into the larger network organization of the cell. Below, we summarize a number of exciting areas that require additional characterization to provide a more complete picture (Figure 6).

AMPK regulation of autophagy

Nutrient-dependent autophagy is best characterized in terms of amino-acid starvation, but autophagy is also activated in response to energy deprivation, for example when mammalian

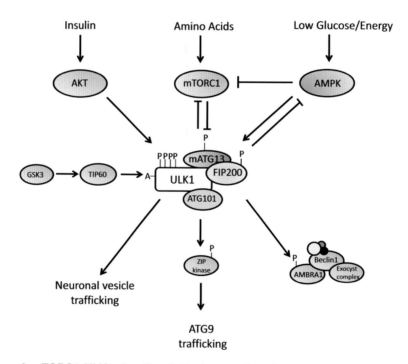

Figure 6. mTORC1–ULK1 signalling in the large cell context
The basic TOR–ATG1/mTORC1–ULK1 signalling axis is now well understood. A number of exciting areas have been identified that require further characterization. Although mTORC1–ULK1 signalling is regulated by amino-acid starvation, a different mechanism linking ULK1 to AMPK in response to energy/glucose deprivation has been observed. Consistent with a tightly regulated nutrient-dependent signalling core, there are multiple feedback loops linking mTORC1, AMPK and ULK1. Growth factor starvation can activate autophagy via the GSK3/Tip60 acetylation pathway. ULK1 has multiple downstream substrates that may add further control to autophagy. AMBRA1 phosphorylation may bridge ULK1 and Beclin1 signalling. ZIP kinase phosphorylation may regulate ATG9 trafficking. The ULK1 complex has a wider role for neuronal vesicle trafficking. A, acetylation.

cells are starved of glucose [22,29,30]. The link of cellular energetics to autophagy highlights a complex signalling network that interconnects AMPK (AMP-activated protein kinase), mTORC1 and ULK1. When energy is limiting, cellular ATP levels decrease while AMP levels increase, leading to an activation of AMPK and phosphorylation of a wide range of downstream targets that control cell energy balance. AMPK plays a key role in the activation of autophagy and this mode of regulation involves phosphorylation of ULK1 on at least six serine and threonine residues within the internal regulatory region (reviewed in [6]). Details on how these AMPK phosphorylation events regulate autophagy need clarification, but there seems to be a mechanism of cross-talk as mTORC1 phosphorylation of ULK1 disrupts the AMPK–ULK1 interaction [22]. These recent insights are consistent with a tight interconnection between mTORC1/AMPK/ULK1 signalling that integrates multiple cell pathways in response to nutrients. As with TOR regulation of autophagy, AMPK-dependent signalling to ATG1 arose early in evolution as supported by evidence from yeast [31].

Multiple mTORC1–AMPK–ULK1 connections

It had already been recognized that AMPK inhibits mTORC1. One route involves AMPK-dependent stimulation of the tuberous sclerosis protein complex TSC2–TSC1, which is a GTPase-activating factor that negatively regulates the GTP-binding protein Rheb. As Rheb is required for mTORC1 activity, low energy and activated AMPK thereby cause TSC2–TSC1 to inactivate Rheb-mTORC1. In the second inhibitory pathway, AMPK directly phosphorylates the mTORC1 subunit raptor (regulatory-associated protein of mTOR). This phosphorylation stimulates binding between raptor and 14-3-3 proteins which results in mTORC1 inhibition. Although AMPK and mTORC1 coordinate to signal downstream of ULK1, there are additional connections to signal upstream. ULK1 can phosphorylate all three subunits of AMPK and repress their activity [32], which may represent a form of negative feedback. Another feedback loop links ATG1 to the inhibition of mTOR as first demonstrated in *Drosophila* [33]. Mammalian cells have a homologous pathway that includes ULK1-dependent phosphorylation of raptor leading to decreased mTORC1 activity [34,35]. As such, AMPK- and ULK1-mediated phosphorylation of raptor work together to inhibit mTORC1 function, but these mechanisms must coordinate with additional raptor phosphorylation events that positively regulate mTORC1. As mTORC1 primarily serves to repress ULK, feedback from ULK1/2 that inhibits mTORC1 might form a positive reinforcement circuit. These multiple mechanisms are likely to enable the cell to finely tune amino acid and energy signals to tightly coordinate autophagy with other homoeostatic pathways.

Post-translational and transcriptional control of ULK1

mTORC1–AMPK–ULK1 have been established as core components of the nutrient-dependent autophagy module. ULK1 regulation has yet other layers of complexity to be discovered. Proteomic MS studies have so far revealed 29 phosphorylation sites in yeast ATG1 and similar approaches have highlighted over 16 phosphorylation sites in ULK1 [6]. Phosphorylation events are observed throughout the different functional regions of ATG1/ULK1 (Figure 5).

These modifications on ATG1/ULK1 reflect combined action of autophosphorylation and other kinases. Autophosphorylation of ATG1 and ULK1 in the kinase domain is critical for catalytic function of the protein and autophagy regulation [36–38]. Future experiments need to characterize more precise roles for the AMPK and mTORC1 sites found in the internal regulatory domain of ULK1. In addition, this region may serve to integrate signals from a range of kinases. There is evidence that protein kinase B (also known as Akt) can also phosphorylate ULK1 within its internal regulatory region [38]. Furthermore, several sites within the ULK1 C-terminal EAT domain are phosphorylated. For these, it is attractive to speculate additional regulatory mechanisms to control membrane binding and protein interactions.

ULK1 serves a central role during early autophagy regulation and so it is reasonable that this component receives multiple upstream signals. Emerging evidence highlights that ULK1 is regulated by protein acetylation on lysine residues in the kinase and internal regulatory regions [39]. Protein acetylation is a form of post-translational modification more widely appreciated for its roles in regulating histones and the p53 tumour suppressor. Mammalian cells also have a pathway where prolonged growth factor starvation causes activation of GSK3 (glycogen synthase kinase 3) which, in turn, stimulates the acetyltransferase Tip60 [HIV-1 Tat (transactivator of transcription)-interactive protein 60 kDa)] to acetylate ULK1 and promote autophagy. Interestingly, acetylation is being recognized for its fundamental role in coordinating cellular metabolic pathways including autophagy [40–42]. Other types of prolonged stress can activate ULK1 function by increasing gene expression. For example, ULK1 expression is increased in a number of cancer cells through the action of p53 and ATF4 (activating transcription factor 4) following DNA damage, hypoxia or disruption of ER homoeostasis (also termed the unfolded protein response) [43,44]. These acetylation and transcriptional mechanisms provide a complementary level of control on ULK1 that function on relatively slower timescales than the amino-acid-sensitive mTORC1/ULK1 pathway.

Downstream of ATG1/ULK1

Fundamental questions still remain on how ATG1/ULK1 truly controls autophagy. The activated ATG1/ULK1 complex translocates on to early autophagy initiation to direct assembly, but the underlying biochemical basis is unclear. Phosphorylation or autophosphorylation on ATG1/ULK1 might drive the assembly by promoting a conformational change and interactions with membranes and additional autophagy factors. Activated ULK1 also phosphorylates the mATG13 and FIP200 subunits of the core complex and although details need to be resolved, these modifications could be serving as autophagy recruitment signals [18,23–25,45]. Besides driving the formation of early autophagy membranes, ULK1 phosphorylates several other substrates and these events may link ATG1/ULK1 with other major autophagy regulatory pathways.

Consistent with a role in controlling intracellular membranes, ATG1/ULK1 signalling directs trafficking of the ATG9 autophagy transmembrane proteins towards autophagosome assembly or elongation and this process is conserved from yeast [46,47]. Experiments that combined *Drosophila* and mammalian systems identified ATG1/ULK1-dependent phosphorylation of a myosin light-chain kinase protein (ZIP kinase in mammals, spaghetti-squash activator in *Drosophila*) [48]. Once activated by ULK1, ZIP kinase phosphorylates myosin II regulatory light-chain to control starvation-induced trafficking through direct interaction with ATG9, providing an additional mechanism for ULK1 to modulate downstream autophagy.

AMBRA1 (activating molecule in BECN1-regulated autophagy 1) is an interesting ULK1 substrate that bridges to Beclin1/VPS34 signalling [49]. ULK1-mediated phosphorylation was required for disassembly of a dynein-light-chain–AMBRA1 complex from a dynein microtubule-dependent motor. Thus it has been proposed that ULK1 might regulate autophagy by allowing release and translocation of an active AMBRA1–Beclin1–VPS34 complex to autophagosome assembly sites. One should note that AMBRA1 has further levels of complexity with mechanisms for regulating autophagy by promoting ubiquitination, protein stability and activity of ULK1 [50]. Additional mechanisms also exist to link the ULK and Beclin1–VPS34 pathways. Components of the exocyst complex (better understood for secretory membrane transport) were shown to directly interact with ULK1 and Beclin1–VPS34 [51]. The model from this work has ULK1 directing formation and localization of different subcomplexes of Beclin1–VPS34 with multiple exocyst subunits. Thus ULK1 has several substrates and binding partners that may allow mechanistic coordination of ULK1, Beclin1 and ATG9 autophagy regulatory pathways.

ATG1/ULK1 for neurobiology

A discussion on ATG1/ULK1 function needs to include roles in neuronal vesicular transport that occur in parallel with autophagy. *Caenorhabditis elegans* with mutation of their ATG1 homologue (unc-51) were originally characterized for uncoordinated behaviour with an underlying axonal defect. Roles for ATG1/ULK1 in neuronal development are conserved in flies and mammals and a number of molecular pathways have been described. The ULK1 C-terminal domain binds syntenin and SynGAP which are proteins implicated in neuronal vesicular trafficking [52,53]. *Drosophila* ATG1 directly phosphorylates factors involved in vesicular transport such as unc-14, unc-76 and VAB-8L [54]. Other data support a neuronal signalling complex containing ULK1/2, TrkA NGF (nerve growth factor) receptor, TRAF6 (tumour necrosis factor receptor-associated factor 6) and p62 [55]. Interestingly, despite the progress made in understanding these molecular details, it still remains unclear how the ATG1/ULK1 complex may be coordinating autophagy and more general vesicular trafficking pathways in neurons and perhaps wider cell types.

Conclusion

Resting cells carry out low-level autophagy as a basal homoeostatic mechanism. Following nutrient starvation, autophagy is rapidly activated to degrade proteins and recycle basic cellular building blocks. The autophagy response on amino-acid depletion is controlled by the conserved TOR–ATG1 signalling pathway. We now understand that activated ATG1 family proteins drive autophagy by localizing to sites of autophagosome formation and assembling complexes of additional autophagy protein factors and membrane precursors. As autophagy plays a universal role in maintaining normal biology of cells and is implicated in multiple disease contexts, it has become important to understand all facets of autophagy regulation in order to devise strategies of modulating the process. ATG1 is an attractive signalling pathway for future study due to its central role in coordinating autophagy and other cellular processes.

Summary

- When amino-acid levels are limiting, the cell responds by inhibiting protein synthesis while increasing protein degradation through autophagy.
- Activation of canonical autophagy following amino-acid starvation is controlled by a conserved pathway centred by TOR and the autophagy kinase ATG1.
- An activated ATG1 protein complex is assembled early during the regulatory process at sites of autophagosome formation.
- Regions within ATG1 have dedicated functions such as binding protein factors, membrane localization and kinase-mediated remodelling of autophagosome assembly.
- Mammalian ULK1 is a signalling nexus that integrates multiple post-translational and gene expression signals to control downstream autophagy and membrane trafficking.

References

1. Yang, Z. and Klionsky, D.J. (2010) Eaten alive: a history of macroautophagy. Nat. Cell Biol. **12**, 814–822
2. Mizushima, N., Levine, B., Cuervo, A.M. and Klionsky, D.J. (2008) Autophagy fights disease through cellular self-digestion. Nature **451**, 1069–1075
3. Rabinowitz, J.D. and White, E. (2010) Autophagy and metabolism. Science **330**, 1344–1348
4. Mizushima, N. and Komatsu, M. (2011) Autophagy: renovation of cells and tissues. Cell **147**, 728–741
5. Mizushima, N., Yoshimori, T. and Ohsumi, Y. (2011) The role of Atg proteins in autophagosome formation. Annu. Rev. Cell Dev. Biol. **27**, 107–132
6. Chan, E.Y. (2012) Regulation and function of uncoordinated–51 like kinase proteins. Antioxid. Redox Signaling **17**, 775–785
7. Codogno, P., Mehrpour, M. and Proikas-Cezanne, T. (2012) Canonical and non-canonical autophagy: variations on a common theme of self-eating? Nat. Rev. Mol. Cell Biol. **13**, 7–12
8. Chan, E.Y. and Tooze, S.A. (2009) Evolution of Atg1 function and regulation. Autophagy **5**, 758–765
9. Loewith, R. and Hall, M.N. (2011) Target of rapamycin (TOR) in nutrient signaling and growth control. Genetics **189**, 1177–1201
10. Kamada, Y., Funakoshi, T., Shintani, T., Nagano, K., Ohsumi, M. and Ohsumi, Y. (2000) Tor-mediated induction of autophagy via an Apg1 protein kinase complex. J. Cell Biol. **150**, 1507–1513
11. Tsukada, M. and Ohsumi, Y. (1993) Isolation and characterization of autophagy-defective mutants of *Saccharomyces cerevisiae*. FEBS Lett. **333**, 169–174
12. Kabeya, Y., Noda, N.N., Fujioka, Y., Suzuki, K., Inagaki, F. and Ohsumi, Y. (2009) Characterization of the Atg17-Atg29-Atg31 complex specifically required for starvation-induced autophagy in *Saccharomyces cerevisiae*. Biochem. Biophys. Res. Commun. **389**, 612–615
13. Kamada, Y., Yoshino, K., Kondo, C., Kawamata, T., Oshiro, N., Yonezawa, K. and Ohsumi, Y. (2010) Tor directly controls the Atg1 kinase complex to regulate autophagy. Mol. Cell. Biol. **30**, 1049–1058
14. Kabeya, Y., Kamada, Y., Baba, M., Takikawa, H., Sasaki, M. and Ohsumi, Y. (2005) Atg17 functions in cooperation with Atg1 and Atg13 in yeast autophagy. Mol. Biol. Cell **16**, 2544–2553

15. Cheong, H., Nair, U., Geng, J. and Klionsky, D.J. (2008) The Atg1 kinase complex is involved in the regulation of protein recruitment to initiate sequestering vesicle formation for nonspecific autophagy in *Saccharomyces cerevisiae*. Mol. Biol. Cell **19**, 668–681
16. Kraft, C., Kijanska, M., Kalie, E., Siergiejuk, E., Lee, S.S., Semplicio, G., Stoffel, I., Brezovich, A., Verma, M., Hansmann, I. et al. (2012) Binding of the Atg1/ULK1 kinase to the ubiquitin-like protein Atg8 regulates autophagy. EMBO J. **31**, 3691–3703
17. Ragusa, M.J., Stanley, R.E. and Hurley, J.H. (2012) Architecture of the Atg17 complex as a scaffold for autophagosome biogenesis. Cell **151**, 1501–1512
18. Chan, E.Y., Longatti, A., McKnight, N.C. and Tooze, S.A. (2009) Kinase-inactivated ULK proteins inhibit autophagy via their conserved C-terminal domains using an Atg13-independent mechanism. Mol. Cell. Biol. **29**, 157–171
19. Nakatogawa, H., Ohbayashi, S., Sakoh-Nakatogawa, M., Kakuta, S., Suzuki, S.W., Kirisako, H., Kondo-Kakuta, C., Noda, N.N., Yamamoto, H. and Ohsumi, Y. (2012) The autophagy-related protein kinase Atg1 interacts with the ubiquitin-like protein Atg8 via the Atg8 family interacting motif to facilitate autophagosome formation. J. Biol. Chem. **287**, 28503–28507
20. Alemu, E.A., Lamark, T., Torgersen, K.M., Birgisdottir, A.B., Larsen, K.B., Jain, A., Olsvik, H., Overvatn, A., Kirkin, V. and Johansen, T. (2012) ATG8 family proteins act as scaffolds for assembly of the ULK complex: sequence requirements for LC3-interacting region (LIR) motifs. J. Biol. Chem. **287**, 39275–39290
21. Shang, L., Chen, S., Du, F., Li, S., Zhao, L. and Wang, X. (2011) Nutrient starvation elicits an acute autophagic response mediated by Ulk1 dephosphorylation and its subsequent dissociation from AMPK. Proc. Natl. Acad. Sci. U.S.A. **108**, 4788–4793
22. Kim, J., Kundu, M., Viollet, B. and Guan, K.L. (2011) AMPK and mTOR regulate autophagy through direct phosphorylation of Ulk1. Nat. Cell Biol. **13**, 132–141
23. Ganley, I.G., Lam, D.H., Wang, J., Ding, X., Chen, S. and Jiang, X. (2009) ULK1.ATG13.FIP200 complex mediates mTOR signaling and is essential for autophagy. J. Biol. Chem. **284**, 12297–12305
24. Jung, C.H., Jun, C.B., Ro, S.H., Kim, Y.M., Otto, N.M., Cao, J., Kundu, M. and Kim, D.H. (2009) ULK-Atg13-FIP200 complexes mediate mTOR signaling to the autophagy machinery. Mol. Biol. Cell **20**, 1992–2003
25. Hosokawa, N., Hara, T., Kaizuka, T., Kishi, C., Takamura, A., Miura, Y., Iemura, S., Natsume, T., Takehana, K., Yamada, N. et al. (2009) Nutrient-dependent mTORC1 association with the ULK1-Atg13-FIP200 complex required for autophagy. Mol. Biol. Cell **20**, 1981–1991
26. Itakura, E. and Mizushima, N. (2010) Characterization of autophagosome formation site by a hierarchical analysis of mammalian Atg proteins. Autophagy **6**, 764–776
27. Hara, T., Takamura, A., Kishi, C., Iemura, S., Natsume, T., Guan, J.L. and Mizushima, N. (2008) FIP200, a ULK-interacting protein, is required for autophagosome formation in mammalian cells. J. Cell Biol. **181**, 497–510
28. McAlpine, F., Williamson, L.E., Tooze, S.A. and Chan, E.Y. (2013) Regulation of nutrient-sensitive autophagy by uncoordinated 51-like kinases 1 and 2. Autophagy **9**, 361–373
29. Kundu, M., Lindsten, T., Yang, C.Y., Wu, J., Zhao, F., Zhang, J., Selak, M.A., Ney, P.A. and Thompson, C.B. (2008) Ulk1 plays a critical role in the autophagic clearance of mitochondria and ribosomes during reticulocyte maturation. Blood **112**, 1493–1502
30. Cheong, H., Lindsten, T., Wu, J., Lu, C. and Thompson, C.B. (2011) Ammonia-induced autophagy is independent of ULK1/ULK2 kinases. Proc. Natl. Acad. Sci. U.S.A. **108**, 11121–11126
31. Wang, Z., Wilson, W.A., Fujino, M.A. and Roach, P.J. (2001) Antagonistic controls of autophagy and glycogen accumulation by Snf1p, the yeast homolog of AMP-activated protein kinase, and the cyclin-dependent kinase Pho85p. Mol. Cell. Biol. **21**, 5742–5752
32. Loffler, A.S., Alers, S., Dieterle, A.M., Keppeler, H., Franz-Wachtel, M., Kundu, M., Campbell, D.G., Wesselborg, S., Alessi, D.R. and Stork, B. (2011) Ulk1-mediated phosphorylation of AMPK constitutes a negative regulatory feedback loop. Autophagy **7**, 696–706

33. Scott, R.C., Juhasz, G. and Neufeld, T.P. (2007) Direct induction of autophagy by Atg1 inhibits cell growth and induces apoptotic cell death. Curr. Biol. **17**, 1–11
34. Lee, S.B., Kim, S., Lee, J., Park, J., Lee, G., Kim, Y., Kim, J.M. and Chung, J. (2007) ATG1, an autophagy regulator, inhibits cell growth by negatively regulating S6 kinase. EMBO Rep. **8**, 360–365
35. Dunlop, E.A., Hunt, D.K., Acosta-Jaquez, H.A., Fingar, D.C. and Tee, A.R. (2011) ULK1 inhibits mTORC1 signaling, promotes multisite raptor phosphorylation and hinders substrate binding. Autophagy **7**, 737–747
36. Yeh, Y.Y., Wrasman, K. and Herman, P.K. (2010) Autophosphorylation within the Atg1 activation loop is required for both kinase activity and the induction of autophagy in *Saccharomyces cerevisiae*. Genetics **185**, 871–882
37. Yeh, Y.Y., Shah, K.H., Chou, C.C., Hsiao, H.H., Wrasman, K.M., Stephan, J.S., Stamatakos, D., Khoo, K.H. and Herman, P.K. (2011) The identification and analysis of phosphorylation sites on the Atg1 protein kinase. Autophagy **7**, 716–726
38. Bach, M., Larance, M., James, D.E. and Ramm, G. (2011) The serine/threonine kinase ULK1 is a target of multiple phosphorylation events. Biochem. J. **440**, 283–291
39. Lin, S.Y., Li, T.Y., Liu, Q., Zhang, C., Li, X., Chen, Y., Zhang, S.M., Lian, G., Ruan, K., Wang, Z. et al. (2012) GSK3-TIP60-ULK1 signaling pathway links growth factor deprivation to autophagy. Science **336**, 477–481
40. Zhao, S., Xu, W., Jiang, W., Yu, W., Lin, Y., Zhang, T., Yao, J., Zhou, L., Zeng, Y., Li, H. et al. (2010) Regulation of cellular metabolism by protein lysine acetylation. Science **327**, 1000–1004
41. Lee, I.H. and Finkel, T. (2009) Regulation of autophagy by the p300 acetyltransferase. J. Biol. Chem. **284**, 6322–6328
42. Yi, C., Ma, M., Ran, L., Zheng, J., Tong, J., Zhu, J., Ma, C., Sun, Y., Zhang, S., Feng, W. et al. (2012) Function and molecular mechanism of acetylation in autophagy regulation. Science **336**, 474–477
43. Pike, L.R., Singleton, D.C., Buffa, F., Abramczyk, O., Phadwal, K., Li, J.L., Simon, A.K., Murray, J.T. and Harris, A.L. (2013) Transcriptional up-regulation of ULK1 by ATF4 contributes to cancer cell survival. Biochem. J. **449**, 389–400
44. Gao, W., Shen, Z., Shang, L. and Wang, X. (2011) Upregulation of human autophagy-initiation kinase ULK1 by tumor suppressor p53 contributes to DNA–damage-induced cell death. Cell Death Differ. **18**, 1598–1607
45. Joo, J.H., Dorsey, F.C., Joshi, A., Hennessy-Walters, K.M., Rose, K.L., McCastlain, K., Zhang, J., Iyengar, R., Jung, C.H., Suen, D.F. et al. (2011) Hsp90–cdc37 chaperone complex regulates ulk1- and atg13-mediated mitophagy. Mol. Cell **43**, 572–585
46. Mari, M., Griffith, J., Rieter, E., Krishnappa, L., Klionsky, D.J. and Reggiori, F. (2010) An Atg9-containing compartment that functions in the early steps of autophagosome biogenesis. J. Cell Biol. **190**, 1005–1022
47. Young, A.R., Chan, E.Y., Hu, X.W., Kochl, R., Crawshaw, S.G., High, S., Hailey, D.W., Lippincott-Schwartz, J. and Tooze, S.A. (2006) Starvation and ULK1-dependent cycling of mammalian Atg9 between the TGN and endosomes. J. Cell Sci. **119**, 3888–3900
48. Tang, H.W., Wang, Y.B., Wang, S.L., Wu, M.H., Lin, S.Y. and Chen, G.C. (2011) Atg1-mediated myosin II activation regulates autophagosome formation during starvation-induced autophagy. EMBO J. **30**, 636–651
49. Di Bartolomeo, S., Corazzari, M., Nazio, F., Oliverio, S., Lisi, G., Antonioli, M., Pagliarini, V., Matteoni, S., Fuoco, C., Giunta, L. et al. (2010) The dynamic interaction of AMBRA1 with the dynein motor complex regulates mammalian autophagy. J. Cell Biol. **191**, 155–168
50. Nazio, F., Strappazzon, F., Antonioli, M., Bielli, P., Cianfanelli, V., Bordi, M., Gretzmeier, C., Dengjel, J., Piacentini, M., Fimia, G.M. and Cecconi, F. (2013) mTOR inhibits autophagy by controlling ULK1 ubiquitylation, self-association and function through AMBRA1 and TRAF6. Nat. Cell Biol. **15**, 406–416

51. Bodemann, B.O., Orvedahl, A., Cheng, T., Ram, R.R., Ou, Y.H., Formstecher, E., Maiti, M., Hazelett, C.C., Wauson, E.M., Balakireva, M. et al. (2011) RalB and the exocyst mediate the cellular starvation response by direct activation of autophagosome assembly. Cell **144**, 253–267
52. Tomoda, T., Kim, J.H., Zhan, C. and Hatten, M.E. (2004) Role of Unc51.1 and its binding partners in CNS axon outgrowth. Genes Dev. **18**, 541–558
53. Mochizuki, H., Toda, H., Ando, M., Kurusu, M., Tomoda, T. and Furukubo-Tokunaga, K. (2011) Unc–51/ATG1 controls axonal and dendritic development via kinesin-mediated vesicle transport in the *Drosophila* brain. PLoS ONE **6**, e19632
54. Toda, H., Mochizuki, H., Flores, 3rd, R., Josowitz, R., Krasieva, T.B., Lamorte, V.J., Suzuki, E., Gindhart, J.G., Furukubo-Tokunaga, K. and Tomoda, T. (2008) UNC-51/ATG1 kinase regulates axonal transport by mediating motor-cargo assembly. Genes Dev. **22**, 3292–3307
55. Zhou, X., Babu, J.R., da Silva, S., Shu, Q., Graef, I.A., Oliver, T., Tomoda, T., Tani, T., Wooten, M.W. and Wang, F. (2007) Unc-51-like kinase 1/2-mediated endocytic processes regulate filopodia extension and branching of sensory axons. Proc. Natl. Acad. Sci. U.S.A. **104**, 5842–5847

2

Omegasomes: PI3P platforms that manufacture autophagosomes

Rebecca Roberts and Nicholas T. Ktistakis[1]

Signalling Programme, Babraham Institute, Cambridge CB22 3AT, U.K.

Abstract

Autophagy is a conserved survival pathway, which cells and tissues will activate during times of stress. It is characterized by the formation of double-membrane vesicles called autophagosomes inside the cytoplasm. The molecular mechanisms and the signalling components involved require specific control to ensure correct activation. The present chapter describes the formation of autophagosomes from within omegasomes, newly identified membrane compartments enriched in PI3P (phosphatidylinositol 3-phosphate) that serve as platforms for the formation of at least some autophagosomes. We discuss the signalling events required to nucleate the formation of omegasomes as well as the protein complexes involved.

Keywords:
Beclin1, omegasome, PI3P, ULK complex, Vps34.

Introduction

Autophagy is a catabolic process in which proteins and organelles are sequestered into double-membrane vesicles for degradation in lysosomes. It is conserved in all organisms, from simple eukaryotes such as yeast to higher eukaryotes such as humans, where it has developed multiple functions. All cells have a basal level of autophagy that has a housekeeping role and helps eliminate harmful material. However, the levels of autophagy can increase dramatically in response

[1] To whom correspondence should be addressed (email nicholas.ktistakis@babraham.ac.uk).

to changes of the cellular environment such as nutrient deprivation or invasion of a pathogen. Three forms of autophagy are described which are termed macroautophagy, microautophagy and chaperone-mediated autophagy. Macroautophagy, commonly referred to as autophagy, is the subject of the present chapter. It involves the formation of double-membrane vesicles known as autophagosomes, which fuse with lysosomes. Autophagy is a survival mechanism which cells activate during times of stress and it is antagonistic to apoptosis.

Autophagy

Much of what is known about the core machinery of autophagy was discovered in yeast, and genes that are involved in this pathway are termed Atg (autophagy-related) [1]. The proteins coded by these genes are organized into distinct complexes that are involved at specific stages of the process. In broad terms, formation of autophagosomes involves an initiation stage, a nucleation stage, an elongation stage and finally closure of the double-membrane structure to form an autophagosome. The master regulator through which many signals converge for the initiation of autophagy is the mTOR (mechanistic target of rapamycin; also known as mammalian target of rapamycin) protein kinase [2–5]. Under nutrient-rich growth conditions, mTOR is active and autophagy is suppressed by the mTOR-induced phosphorylation of Atg1, the only kinase among the Atg proteins. mTOR can form two complexes (mTORC1 and mTORC2), which differ in sensitivity to rapamycin and in subunit composition. The subunits in each of these complexes that determine the sensitivity to rapamycin are known as raptor (regulatory-associated protein of mTOR) for mTORC1 and rictor (rapamycin-insensitive companion of mTOR) for mTORC2. mTORC1 is the complex that directly regulates the autophagic pathway. Upon starvation, or in response to other stimuli, mTORC1 is inactivated which rapidly results in dephosphorylation of Atg1 and the induction of autophagy. In mammalian cells, there are two Atg1 homologues known as ULK (uncoordinated-51-like kinase) 1 and ULK2. There is some redundancy between the two ULKs with ULK1 having the main role [6]. ULK is one of the earliest proteins activated during the induction of autophagy, and is in complex with Atg13, Atg101 and FIP200 (focal adhesion kinase family-interacting protein 200 kDa) (the ULK complex). This ULK complex is essential for autophagosome formation and is thought to interact directly with mTOR under nutrient-rich conditions. During starvation, the ULK complex dissociates from mTOR and can be recruited to the site of autophagosome formation [3] (Figure 1). Concomitant to the physical disengagement from mTOR, a series of phosphorylation/dephosphorylation events involving ULK1, Atg13 and FIP200 ensures that the complex arriving at the autophagosome formation site is competent for the downstream steps. Recently, it has been shown that AMPK (AMP-activated protein kinase) can also play a role in the regulation of ULK during autophagy. The evidence implies that AMPK directly interacts with the ULK complex and phosphorylates several sites important for ULK activation. This is enhanced by mTOR inhibitors such as rapamycin, indicating that mTORC1 negatively regulates this association. Regulation of autophagy by both mTORC1 and AMPK allows the cells to respond to a wide range of stimuli [7,8].

Autophagosomes in yeast form at a site known as the PAS (phagophore assembly site), whereas in mammalian cells there are multiple such sites at any one time [4]. Once the ULK complex is at the PAS, it helps recruit the next complex in the pathway, namely the Vps34 complex I. Figure 1 highlights the key components required for the formation of the autophagosomes, and involvement of the ULK complex and Vps34 complex I.

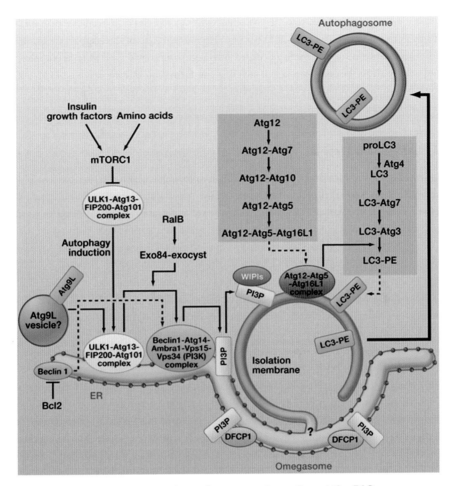

Figure 1. Events in omegasome/autophagosome formation at the PAS
Autophagy is negatively regulated by mTORC1. During starvation, the ULK complex translocates to the ER. The formation of omegasomes at sites on the ER requires the recruitment of the ULK complex (consisting of ULK1, Atg13, FIP200 and Atg101) and the Vps34 complex I (consisting of Beclin1, Vps34, Vps15 and Ambra-1/Atg14). These are required for PI3P generation which localizes the next proteins required for the progression of the pathway such as WIPI2 and DFCP1. This allows further recruitment of autophagy machinery such as the Atg12–Atg5–Atg16 complex, and LC3–PE in order to generate a complete double-membrane autophagosome from the isolation membrane. The Atg12–Atg5–Atg16 complex and Atg9 are essential for the progression of the autophagosome, and are required for LC3 localization. Reprinted from [1] with permission from Elsevier, copyright (2011) [Mizushima, N. and Komatsu, M. (2011) Autophagy: renovation of cells and tissues. Cell **147**, 728–741].

The Vps34 complex I consists of Atg14L, Beclin1 (Atg6), Vps34 [class III PI3K (phosphoinositide 3-kinase)] and Vps15. The lipid kinase Vps34 is crucial in producing the lipid signalling molecule PI3P (phosphatidylinositol 3-phosphate) resulting in recruitment of other Atg proteins and allowing for expansion of the autophagosomal membrane [1,2,5]. Membrane expansion is mediated by the actions of the Atg5–Atg12–Atg16 complex, and by vesicles containing Atg9 and cycling between the PAS and organelles. The Atg5–Atg12–Atg16 complex mediates the conjugation of Atg8, known in mammalian cells as MAP1LC3 (microtubule-associated protein 1 light-chain 3) or simply LC3, with PE (phosphatidylethanolamine) as it becomes incorporated into the forming autophagosomes. LC3 is a very abundant protein of the autophagosome inner and outer membrane and remains associated with autophagosomes until their degradation in lysosomes [3,5]. As the autophagosomal membrane expands and begins to close on itself, it also sequesters within it cytoplasmic material (proteins or whole organelles) destined for delivery to lysosomes for degradation.

Omegasomes

Many EM (electron microscopy) studies looking at autophagosome formation concluded that the membranes must be related to the ER (endoplasmic reticulum) [9,10], but clear ER-derived markers that become part of the autophagosomal membrane are lacking. Previous work from our group proposed the idea that specialized membrane compartments linked to the ER and enriched in PI3P (termed omegasomes due to their Ω shape) may be sites of autophagosome formation [11]. A novel PI3P-binding protein named DFCP1 (double FYVE domain-containing protein 1) was used to illustrate the relationship between the ER, omegasomes and autophagosome formation. DFCP1 has an ER targeting domain and two FYVE domains, localizing to the ER and Golgi, instead of endosomes [12] (Figure 2A). During starvation, DFCP1 translocates from the cytosol to distinct puncta throughout the cytoplasm (Figure 2B) which partially colocalize with LC3 and Atg5 and are identified as sites of autophagosome formation. The translocation of DFCP1 could be blocked by PI3K inhibitors such as wortmannin or 3-methyladenine, or by using siRNA to knock down levels of Vps34 and Beclin1, indicating that it requires formation of PI3P. It is also apparent that omegasome expansion is seen in close apposition to Vps34-containing vesicles, providing a potential way for the delivery of the PI3P-synthesizing machinery to these sites (Figure 2C). These DFCP1-containing puncta are clearly connected to the ER and their dynamic behaviour during autophagy is reflected in changes to the underlying ER structure. Figure 2(D) illustrates that as omegasomes form and expand (green colour) the underlying ER also expands (blue colour) leading to an almost complete colocalization. In terms of the omegasome and LC3 dynamics, it is apparent that LC3 puncta (red colour) form within omegasomes and at the point of maximal expansion the LC3 membranes exit omegasomes accompanied by gradual diminishment of the omegasome signal. The LC3 particle exiting omegasomes requires 3–5 min to mature and become acidic, supporting the idea that autophagosomes are not mature when leaving the omegasome. The connection between expanding omegasomes and the underlying ER is perhaps more clearly seen by TIRF (total internal reflection fluorescence) microscopy (Figure 2E).

The idea that the ER is one site of autophagosome formation is supported by previous electron tomography studies showing many examples of the ER ribbon weaving through the autophagosome membrane [13,14]. Other data in the literature imply mitochondria as a source for autophagosome membrane, suggesting this organelle contributes as much as the ER membrane to autophagosome formation [15]. Recent work by Hamasaki et al. [16] indicates that the ER–mitochondria contact sites appear to label with early autophagy proteins during starvation and disruption of these contact sites results in a decrease in autophagy. Thus it is possible that autophagosomes are also induced at the junction of the ER and mitochondria [16]. Similarly, inhibition of endocytosis has a detrimental effect on autophagy, reducing the number of autophagosomes, suggesting the plasma membrane may provide

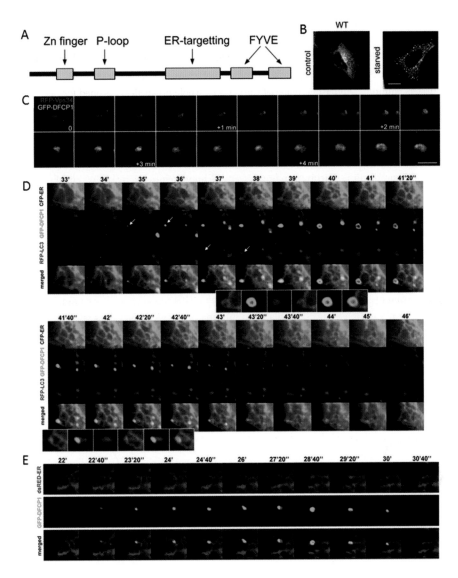

Figure 2. Characterization of DFCP1: a marker for PI3P production and omegasome formation

(**A**) Domain organization for the protein DFCP1 including two FYVE domains that bind the PI3P generated during omegasome induction. (**B**) Localization of DFCP1 in normal medium (diffuse throughout the cell) and during starvation (in small puncta which correspond to omegasomes). (**C**) Localization of GFP–DFCP1 puncta (green) near to RFP–Vps34 puncta (red) indicating a pathway of PI3P generation in the vicinity of omegasomes. (**D**) An example of live cell imaging whereby the formation of omegasomes (followed by GFP–DFCP1) is accompanied by a change in the underlying ER (CFP–ER) and results in the formation of an autophagosome (RFP–LC3) over a period of 40 min. (**E**) Relationship of omegasomes to the ER using TIRF imaging. Adapted from [11] with permission ©Axe, E.L. et al., 2008 (originally published in Axe, E.L., Walker, S.A., Manifava, M., Chandra, P., Roderick, H.L., Habermann, A., Griffiths, G. and Ktistakis, N.T. (2008) Autophagosome formation from membrane compartments enriched in PI(3)P and dynamically connected to the endoplasmic reticulum. J. Cell Biol. **182**, 685–701).

membrane for autophagosomes [17]. The contribution of membranes from several organelles to the phagophore need not be mutually exclusive; it is possible that, depending on the conditions and the stimulus, many cellular membranes may contribute to the formation of autophagosomes [3].

Regulation of omegasome formation

The generation of the lipid PI3P is an important factor for the initiation of autophagy and for localizing effector proteins to the site of autophagosome formation. The most prominent PI3P-binding protein is Atg18. In mammals, Atg18 homologues are called WIPI (WD-repeat protein interacting with phosphoinositides) family members including WIPI1, WIPI2 and WIPI4, all of which have been shown to be involved in autophagy [18–20]. Similar to DFCP1, WIPI2 is recruited to the omegasomes during PI3P production, and is required for LC3 lipidation and closure of the double-membrane structure. In cells down-regulated for WIPI2, omegasomes marked with DFCP1 are enhanced even under basal conditions, indicating that WIPI2 is downstream of DFCP1 and is involved in the maturation of the autophagosome as it leaves the omegasome [18]. A similar result holds for WIPI4 [20].

The Vps34 complex can be regulated in multiple ways and this is illustrated in Figure 3 [21,22]. One of the key proteins of the complex is Beclin1 with an essential role in

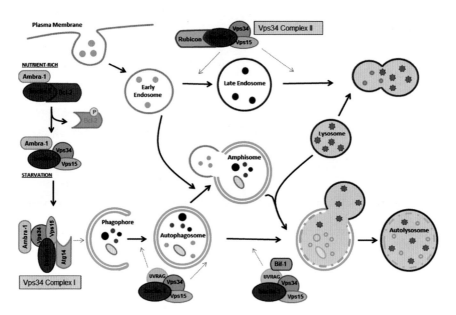

Figure 3. Several Vps34 complexes regulating autophagy or endosomal traffic
Vps34 complex II is involved in the regulation of endocytosis, whereas Vps34 complex I is important for the induction of autophagy. Under nutrient-rich conditions Beclin1 binds to Bcl-2 family members; during starvation Bcl-2 proteins become phosphorylated and release Beclin1 to form part of the Vps34 type I complex. This complex can be further activated by Ambra-1 and/or Atg14 allowing autophagy to occur. Other binding partners of complex I include Bif-1 and UVRAG. Autophagosomes may fuse with endocytic vesicles termed amphisomes, and can be delivered to lysosomes for degradation. For the regulation of endocytosis, the Vps34 complex II may bind to Rubicon.

autophagy; as much as 50% of cellular Beclin1 is in complex with Vps34 and when overexpressed can lead to certain forms of cancer [23]. Under nutrient-rich conditions, autophagy is inhibited due to Beclin1 binding to Bcl-2 family members [24]. However, during starvation, the Bcl-2 proteins can become phosphorylated and no longer bind Beclin1, allowing autophagy to occur. Once Beclin1 is free from Bcl-2 interactions, it can bind to other proteins such as UVRAG (UV radiation resistance-associated gene), Ambra1, Atg14L or Rubicon to regulate autophagy [22]. UVRAG and Atg14L binding to the PI3K complex are mutually exclusive and evidence suggests that both are involved in the early stages of autophagosome formation. Atg14 can localize to the omegasomes without binding to the PI3K complex, and it is thought that its role is to recruit the complex to the omegasome and stimulate production of PI3P [25]. UVRAG increases the interaction between Beclin1 and Vps34, thus promoting autophagy. Similarly, downstream of these early events, Bif-1 binds to UVRAG rather than Beclin1, and can also be a positive inducer of autophagy [26,27]. On the other hand, Rubicon has been shown to inhibit autophagy. Rubicon has been shown to bind to the Beclin1 complexes containing UVRAG and inhibits autophagosome maturation by inhibiting the kinase activity of Vps34, and inhibiting degradation in lysosomes [28].

In addition to indirect effects via Beclin1, the formation of PI3P during omegasome nucleation and expansion could be regulated by direct effects on Vps34 itself. Previous studies indicate that at least three kinases phosphorylate Vps34 including cyclin-dependent kinase 1, cyclin-dependent kinase 5 and protein kinase D. Phosphorylation by the first two results in inhibition, whereas protein kinase D phosphorylation has been shown to increase Vps34 activity [29,30]. Undoubtedly more proteins will be identified which can regulate Vps34 directly. The challenge will be to differentiate effects on the endocytic function of Vps34 (mediated by complex II) versus the autophagic function (mediated by complex I).

Omegasomes during pathogen-induced autophagy

Autophagy has a complicated relationship with pathogen infection. In some cases pathogens usurp the autophagic machinery for replication and expansion, whereas in other cases pathogens activate autophagy during infection and this can be part of an innate immune response [31]. Beclin1 is a key player in autophagy and during viral infection appears to be the common target protein in several instances. For example, the herpes simplex virus protein ICP34.5 binds to Beclin1 to prevent activation of autophagy and clearance of infection [32]. A screen of non-structural proteins from infectious bronchitis virus revealed that one of them (nsp6) was capable of inducing omegasomes that led to autophagosome formation. The equivalent proteins from mammalian coronaviruses and porcine reproductive and respiratory syndrome viruses, in all cases multi-spanning ER transmembrane proteins, were all capable of inducing omegasomes leading to autophagosomes [33]. Similarly, *Salmonella* infection has been shown to trigger formation of omegasomes before their maturation into autophagosomes in a pathway also requiring the small GTPase Rab1, a known regulator of ER-to-Golgi traffic [34]. Thus in the case of both virally and bacterially induced autophagy it appears that an omegasome intermediate is used.

Termination of the PI3P signal: role for several 3-phosphatases

Lipid signalling depends not only on the generation of the signal (which is usually rapid and localized), but also on its consumption. In the case of PI3P during autophagy, several 3-phosphatases have been shown to be involved, with none so far being essential on its own. These phosphatases all belong to the MTM (myotubularin) family, they are ubiquitously expressed and are known to play a role in the endocytic pathway. A recent siRNA screen has revealed that three such phosphatases play a role in the regulation of autophagy, Jumpy (MTMR14), MTMR6 and MTMR7, with Jumpy having a stronger role [35,36], whereas another group implicated MTMR3 in the termination of the signal during autophagosome formation [34]. Knockdown of the expression of both MTMR3 and Jumpy increased autophagosome number and both proteins localized to the phagophore membrane [36,37]. One possibility to keep in mind is that multiple pools of PI3P may be involved in autophagosome formation, one during the induction step related to omegasomes and one at later steps that could mediate maturation and/or fusion with the lysosomes. It will be important in future studies to differentiate between these steps in order to provide additional insights into the specificity of these 3-phosphatases during autophagosome formation.

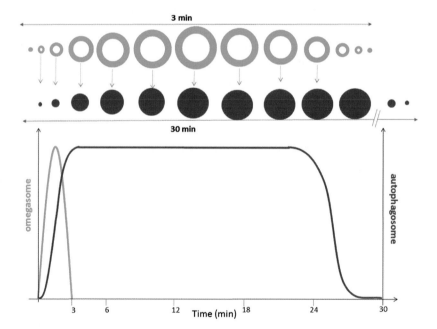

Figure 4. Omegasomes and autophagosome lifetime
A view of the relative lifetime of omegasomes (green rings) and autophagosomes (red circles) that form during induction of autophagy. Omegasomes are transient precursors that typically last for 3 min, whereas autophagosomes can last up to 30 min. The graph illustrates the sharp increase and decline in omegasomes in comparison with the long-lasting lifetime of autophagosomes, which are eventually degraded in lysosomes. Reprinted from [38] with permission from Landes Bioscience Copyright (2011) [Ktistakis, N.T., Andrews, S. and Long, J. (2011) What is the advantage of a transient precursor in autophagosome biogenesis. Autophagy **7**, 118–122].

Conclusion

Part of the requirement for PI3P formation during the induction of autophagy can be understood in the context of omegasome action. These omegasomes are transient precursors from which autophagosomes form. The advantage of such a transient precursor appears to be twofold [38] (Figure 4). It allows the cells to be able to respond to a wide range of nutrient states, and it also makes for a smooth and economical approach to a new steady state of autophagosome formation in response to a starvation signal. However, it should be noted that autophagy appears to be regulated in many ways and at multiple points, and the pathway involving omegasome intermediates is one of several leading to autophagosome formation. Among the many unknowns in the field, perhaps the plasticity of the autophagic response and the underlying rationale for such plasticity will be important questions for future studies.

Summary

- Autophagy is a conserved catabolic process in which proteins and organelles are sequestered into double-membrane vesicles (termed autophagosomes) for degradation in lysosomes.
- Formation of autophagosomes is the decisive step during autophagy, and many proteins (generally termed Atg from work in yeast) are involved in the process.
- Induction of autophagosomes is regulated by the protein kinase ULK and its associated proteins, and by the lipid kinase Vps34 and its associated proteins.
- One function of the Vps34 complex during autophagy is the generation of PI3P-enriched and ER-connected membrane platforms, termed omegasomes, within which autophagosomes are formed.
- Omegasome intermediates are also evident during viral or bacterial-induced autophagy.
- The current view in the field is that autophagosome formation may involve multiple membranes sources (mitochondria, Golgi and plasma membrane) depending on cellular state.

Our work is supported by the Biotechnology and Biological Sciences Research Council.

References

1. Mizushima, N. and Komatsu, M. (2011) Autophagy: renovation of cells and tissues. Cell **147**, 728–741
2. Weidberg, H., Shvets, E. and Elazar, Z. (2011) Biogenesis and cargo selectivity of autophagosomes. Annu. Rev. Biochem. **80**, 125–156
3. Mizushima, N., Yoshimori, T. and Ohsumi, Y. (2012) The role of Atg proteins in autophagosome formation. Annu. Rev. Cell Dev. Biol. **27**, 107–132
4. Rubinsztein, D.C., Shpilka, T. and Elazar, Z. (2012) Mechanisms of autophagosome biogenesis. Curr. Biol. **22**, R29–R34

5. Kuballa, P., Nolte, W.M., Castoreno, A.B. and Xavier, R.J. (2012) Autophagy and the immune system. Annu. Rev. Immunol. **30**, 611–646
6. McAlpine, F., Williamson, L.E., Tooze, S.A. and Chan, E.Y. (2013) Regulation of nutrient-sensitive autophagy by uncoordinated 51-like kinases 1 and 2. Autophagy **9**, 361–373
7. Kim, J., Kundu, M., Viollet, B. and Guan, K.L. (2011) AMPK and mTOR regulate autophagy through direct phosphorylation of Ulk1. Nat. Cell Biol. **13**, 132–141
8. Egan, D.F., Shackelford, D.B., Mihaylova, M.M., Gelino, S., Kohnz, R.A., Mair, W., Vasquez, D.S., Joshi, A., Gwinn, D.M., Taylor, R. et al. (2011) Phosphorylation of ULK1 (hATG1) by AMP-activated protein kinase connects energy sensing to mitophagy. Science **331**, 456–461
9. Dunn, W. (1990) Studies on the mechanisms of autophagy: formation of the autophagic vacuole. J. Cell Biol. **110**, 1923–1933
10. Ueno, T., Muno, D. and Kominami, E. (1991) Membrane markers of endoplasmic reticulum preserved in autophagic vacuolar membranes isolated from leupeptin-administered rat liver. J. Biol. Chem. **266**, 18995–18999
11. Axe, E.L., Walker, S.A., Manifava, M., Chandra, P., Roderick, H.L., Habermann, A., Griffiths, G. and Ktistakis, N.T. (2008) Autophagosome formation from membrane compartments enriched in phosphatidylinositol 3-phosphate and dynamically connected to the endoplasmic reticulum. J. Cell Biol. **182**, 685–701
12. Ridley, S.H., Ktistakis, N., Davidson, K., Anderson, K.E., Manifava, M., Ellson, C.D., Lipp, P., Bootman, M., Coadwell, J., Nazarian, A. et al. (2001) FENS-1 and DFCP1 are FYVE domain-containing proteins with distinct functions in the endosomal and Golgi compartments. J. Cell Sci. **114**, 3991–4000
13. Hayashi-Nishino, M., Fujita, N., Noda, T., Yamaguchi, A., Yoshimori, T. and Yamamoto, A. (2010) Electron tomography reveals the endoplasmic reticulum as a membrane source for autophagosome formation. Autophagy **6**, 301–303
14. Hailey, D.W., Rambold, A.S., Satpute-Krishnan, P., Mitra, K., Sougrat, R., Kim, P.K. and Lippincott-Schwartz, J. (2010) Mitochondria supply membranes for autophagosome biogenesis during starvation. Cell **141**, 656–667
15. Ylä-Anttila, P., Vihinen, H., Jokitalo, E. and Eskelinen, E.L. (2009) 3D tomography reveals connections between the phagophore and endoplasmic reticulum. Autophagy **5**, 1180–1185
16. Hamasaki, M., Furuta, N., Matsuda, A., Nezu, A., Yamamoto, A., Fujita, N., Oomori, H., Noda, T., Haraguchi, T., Hiraoka, Y. et al. (2013) Autophagosomes form at ER–mitochondria contact sites. Nature **495**, 389–393
17. Ravikumar, B., Moreau, K., Jahreiss, L., Puri, C. and Rubinsztein, D.C. (2010) Plasma membrane contributes to the formation of pre-autophagosomal structures. Nat. Cell Biol. **12**, 747–757
18. Proikas-Cezanne, T., Ruckerbauer, S., Stierhof, Y.D., Berg, C. and Nordheim, A. (2007) Human WIPI-1 puncta-formation: a novel assay to assess mammalian autophagy. FEBS Lett. **581**, 3396–3404
19. Polson, H.E., de Lartigue, J., Rigden, D.J., Reedijk, M., Urbé, S., Clague, M.J. and Tooze, S.A. (2010) Mammalian Atg18 (WIPI2) localizes to omegasome-anchored phagophores and positively regulates LC3 lipidation. Autophagy **6**, 506–522
20. Proikas-Cezanne, T., Waddell, S., Gaugel, A., Frickey, T., Lupas, A. and Nordheim, A. (2004) WIPI-1α (WIPI49), a member of the novel 7-bladed WIPI protein family, is aberrantly expressed in human cancer and is linked to starvation-induced autophagy. Oncogene **23**, 9314–9325
21. Burman, C. and Ktistakis, N. (2010) Autophagosome formation in mammalian cells. Semin. Immunopathol. **32**, 397–413
22. He, C. and Levine, B. (2010) The Beclin 1 interactome. Curr. Opin. Cell Biol. **22**, 140–149
23. Abrahamsen, H., Stenmark, H. and Platta, H.W. (2012) Ubiquitination and phosphorylation of Beclin 1 and its binding partners: tuning class III phosphatidylinositol 3-kinase activity and tumor suppression. FEBS Lett. **586**, 1584–1591

24. Liang, X.H., Jackson, S., Seaman, M., Brown, K., Kempkes, B., Hibshoosh, H. and Levine, B. (1999) Induction of autophagy and inhibition of tumorigenesis by Beclin 1. Nature **402**, 672–676
25. Matsunaga, K., Morita, E., Saitoh, T., Akira, S., Ktistakis, N.T., Izumi, T., Noda, T. and Yoshimori, T. (2010) Autophagy requires endoplasmic reticulum targeting of the PI3-kinase complex via Atg14L. J. Cell Biol. **190**, 511–521
26. Liang, C., Feng, P., Ku, B., Dotan, I., Canaani, D., Oh, B.H. and Jung, J.U. (2006) Autophagic and tumour suppressor activity of a novel Beclin1-binding protein UVRAG. Nat. Cell Biol. **8**, 688–699
27. Takahashi, Y., Coppola, D., Matsushita, N., Cualing, H.D., Sun, M., Sato, Y., Liang, C., Jung, J.U., Cheng, J.Q., Mulé, J.J. et al. (2007) Bif-1 interacts with Beclin 1 through UVRAG and regulates autophagy and tumorigenesis. Nat. Cell Biol. **9**, 1142–1151
28. Zhong, Y., Wang, Q.J., Li, X., Yan, Y., Backer, J.M., Chait, B.T., Heintz, N. and Yue, Z. (2009) Distinct regulation of autophagic activity by Atg14L and Rubicon associated with Beclin 1-phosphatidylinositol-3-kinase complex. Nat. Cell Biol. **11**, 468–476
29. Furuya, T., Kim, M., Lipinski, M., Li, J., Kim, D., Lu, T., Shen, Y., Rameh, L., Yankner, B., Tsai, L.H. and Yuan, J. (2010) Negative regulation of Vps34 by Cdk mediated phosphorylation. Mol. Cell **38**, 500–511
30. Eisenberg-Lerner, A. and Kimchi, A. (2012) PKD is a kinase of Vps34 that mediates ROS-induced autophagy downstream of DAPk. Cell Death Differ. **19**, 788–797
31. Levine, B., Mizushima, N. and Virgin, H.W. (2011) Autophagy in immunity and inflammation. Nature **469**, 323–335
32. Orvedahl, A., Alexander, D., Tallóczy, Z., Sun, Q., Wei, Y., Zhang, W., Burns, D., Leib, D.A. and Levine, B. (2007) HSV-1 ICP34.5 confers neurovirulence by targeting the Beclin 1 autophagy protein. Cell Host Microbe **1**, 23–35
33. Cottam, E.M., Maier, H.J., Manifava, M., Vaux, L.C., Chandra-Schoenfelder, P., Gerner, W., Britton, P., Ktistakis, N.T. and Wileman, T. (2011) Coronavirus nsp6 proteins generate autophagosomes from the endoplasmic reticulum via an omegasome intermediate. Autophagy **7**, 1335–1347
34. Huang, J., Birmingham, C.L., Shahnazari, S., Shiu, J., Zheng, Y.T., Smith, A.C., Campellone, K.G., Heo, W.D., Gruenheid, S., Meyer, T. et al. (2011) Antibacterial autophagy occurs at PI(3)P-enriched domains of the endoplasmic reticulum and requires Rab1 GTPase. Autophagy **7**, 17–26
35. Vergne, I. and Deretic, V. (2010) The role of PI3P phosphatases in the regulation of autophagy. FEBS Lett. **584**, 1313–1318
36. Vergne, I., Roberts, E., Elmaoued, R.A., Tosch, V., Delgado, M.A., Proikas-Cezanne, T., Laporte, J. and Deretic, V. (2009) Control of autophagy initiation by phosphoinositide 3-phosphatase Jumpy. EMBO J. **28**, 2244–2258
37. Taguchi-Atarashi, N., Hamasaki, M., Matsunaga, K., Omori, H., Ktistakis, N.T., Yoshimori, T. and Noda, T. (2010) Modulation of local PtdIns3P levels by the PI phosphatase MTMR3 regulates constitutive autophagy. Traffic **11**, 468–478
38. Ktistakis, N.T., Andrews, S. and Long, J. (2011) What is the advantage of a transient precursor in autophagosome biogenesis. Autophagy **7**, 118–122

Current views on the source of the autophagosome membrane

Sharon A. Tooze[1]

London Research Institute, Cancer Research UK, 44 Lincoln's Inn Fields, London WC2A 3LY, U.K.

Abstract

Autophagy was discovered in the late 1950s when scientists using the first electron microscopes saw membrane-bound structures in cells that contained cytoplasmic organelles, including mitochondria. Pursuant to further morphological characterization it was recognized that these vesicles, now called autophagosomes, are found in all eukaryotic cells and undergo changes in morphology from a double-membraned vesicle with recognizable content, i.e. sequestered organelles, to a uniformly dense core autolysosome. Genetic screens in the yeast *Saccharomyces cerevisiae* in the 1990s provided a molecule framework for the next era of discovery during which the interest in, and research into, autophagy has rapidly expanded into many areas of human biology and disease. A relatively small cohort of approximately 36 proteins, called Atgs (autophagy-related proteins), orchestrate the formation of the autophagosome, and these are now being studied and functionally characterized. Although the function of these proteins is being elucidated, the underlying molecular mechanisms of how autophagosomes form are still not completely understood. Recent advances have, however, provided a significant advance in both our understanding of the molecular control of the Atg proteins and the source of the membranes. A consensus view is emerging from these advances that the endoplasmic reticulum is the nucleation site for the autophagosome, and that contributions from other compartments (Golgi, endosomes and plasma membrane) are required. In the present chapter, I review the data from the pre-molecular decades, and discuss the most recent publications to give an overview of the current view of where, and how, autophagosomes form in mammalian cells.

[1]email sharon.tooze@cancer.org.uk

Keywords:
autophagosome, MAM, omegasome, phagophore, PAS.

Introduction

In the 1950s after the development of EM (electron microscopy), cell biologists using newly developed techniques and EM recognized that there was an unusual type of lysosome in some cells which contained mitochondria and cytosolic components [1], subsequently called cytolysosomes [2] or autophagosomes [3]. Later, some of the vesicles were confirmed as *bona fide* lysosomes; use of cytochemical techniques confirmed they contained lysosomal hydrolases. However, what intrigued these scientists (and continues to intrigue us) are the membrane dynamics and topological issues relating to the formation of the double membrane, and the mechanism of internalization of cytosolic components. Both of these issues reinforce the notion that the origin and formation of these vesicles is unique.

The current view of the autophagy pathway is shown in Figure 1. The pathway is initiated at the PAS (pre-autophagosomal structure, also known as phagophore assembly site). It is worth noting from the outset that it is not clear from either morphological or biochemical techniques what the PAS is. A double-membrane structure, seen by EM in cells as an open cup-like structure, forms from the PAS, which is called a phagophore (or isolation membrane). This structure closes to form an autophagosome, which after its closure and formation has a double membrane. One face of this double membrane (which was originally cytosolic) is now within the lumen of the autophagosome. The autophagosome then fuses with endosomes and lysosomes, and acquires degradative enzymes, including proteases and lipases, becoming an autolysosome.

Although the static images collected by EM provide detailed morphological data, they did not reveal the dynamics of the process, in particular where and how rapidly the

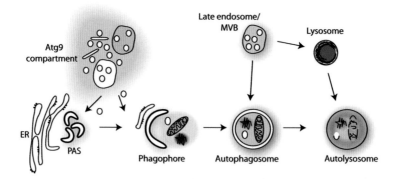

Figure 1. Autophagy pathway in mammalian cells
PASs form in connection with ER membranes. Phagophores, also called isolation membranes, grow, engulf cytoplasmic material and organelles, and close to become autophagosomes. Vesicles containing Atg9 traffic to and from the forming and expanding PAS, phagophore and autophagosome. Autophagosomes then fuse with late endosomes, multivesicular bodies and lysosomes to become autolysosomes.

autophagosomes form, although early EM data did suggest autophagosomes were relatively short-lived [4]. Live-cell imaging techniques [5–8] revealed the rapid kinetics of autophagosome formation, in particular after amino-acid starvation. Autophagosomes typically form within 5–10 min of starvation, which suggests that the formation occurs from pre-assembled structures or by efficient vesicular trafficking from existing compartments and membranes. Thus the following questions arise: what is the PAS, what are phagophores and where do the membranes come from? Finally, how do they become autophagosomes?

Definition of the PAS and phagophore

A key observation, again using EM techniques, was that the phagophore membranes were electron-dense, and more osmophilic than most other cellular membranes [9]. Further morphological advances and high-quality cytochemical studies led to several different hypotheses about the origin of the autophagic membranes, including the possibility that the membrane originated from the ER (endoplasmic reticulum) [10–12] (and references therein) and the GERL (Golgi–ER–lysosome) [13]. Essentially, an electron-dense reaction product produced by the enzymes resident in organelles, such as the ER, Golgi and lysosomes, can be detected by conventional EM in the positive organelles, helping to understand the origin or composition of the compartment. These morphological approaches were expanded by the application of subcellular fractionation techniques, especially powerful when combined with morphology, which aimed to understand what the intermediates in the pathway were, and most crucially, what the composition of the phagophore, or isolation membrane and amphisome were [12,14]. In addition, using freeze-fracture microscopy the phagophore membrane appeared to be protein-poor, suggesting that the membrane was a unique composition [15]. However, technical limitations remained with different laboratories reporting different amounts of enzyme labelling and activity for resident enzymes, or different membrane markers in subcellular fractionation, so, despite the use of these state-of-the-art approaches, no universal consensus was obtained about the origin of the PAS or phagophore.

The molecular machinery

There are now more than 36 known Atg proteins in yeast, and among these, there is a set of proteins that are required for macroautophagy in yeast and mammals. Macroautophagy, also known as autophagy, is the best studied type of autophagy, among the numerous types that include: selective autophagy (mitophagy and xenophagy); microautophagy; and the mammalian-specific chaperone-mediated autophagy. (Macro)autophagy is thought to be a non-selective process, although some of the cargo-selection machinery used in selective autophagy is also required (see Chapter 5). In the case of the yeast proteins there is typically a single gene and protein, whereas in mammals there is some gene duplication. Both the yeast and mammalian proteins can be grouped into functional categories, and arranged in a hierarchy (Figure 2). Briefly, in yeast, the Atg1 serine–threonine kinase complex, including Atg1, Atg13, Atg17, Atg29 and Atg31, is thought to be the most upstream regulator of autophagy, and downstream of TOR (target of rapamycin). In mammals, this complex is the ULK (uncoordinated-51-like kinase) complex comprising ULK1/2, FIP200 (focal adhesion kinase family-interacting protein 200 kDa) and Atg13, and is also regulated by mTORC1 [mammalian (also

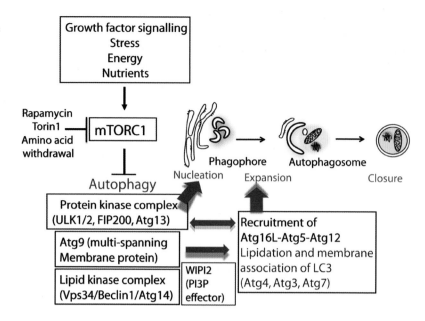

Figure 2. Molecular signalling pathway and hierarchy of Atg proteins in mammalian starvation-induced autophagy
Autophagy is initiated by alterations in growth conditions, and local energy supply. These changes are sensed by the master regulators mTORC1 and AMPK (not shown). mTORC1 is a negative regulator of autophagy and acts upstream of the ULK1/2 kinase complex. The hierarchy of the Atg proteins downstream of the initiating inactivation of mTORC1 and activation of ULK1/2 complex is being established, and likely will also involve direct action as well as some cross-talk.

known as mechanistic) TOR complex 1] and AMPK (AMP-activated protein kinase) (see Chapter 1). The second conserved kinase complex is a lipid kinase complex containing Vps34, and is also known as the class III PI3K (phosphoinositide 3-kinase) complex in mammals. The autophagy-specific PI3K complex consists of a catalytic subunit, Vps34, the regulatory subunit p150 (homologue of yeast Vps15), Beclin1 (in yeast, Vps30) and Atg14. Using inhibitors of the PI3Ks, it was shown that PI3P (phosphatidylinositol 3-phosphate) is essential for the initiation of autophagy. It is not known what role PI3P has in the organization of the phagophore or autophagosome membrane. However, as is also the case in other intracellular trafficking steps, the effectors of PI3P are thought to have important functions [16]. Effectors of PI3P produced during autophagy are proposed to be the WIPI (WD-repeat protein interacting with phosphoinositides) family (WIPI1–4), which are the homologues of yeast Atg18 and Atg21. An additional effector of the pool of PI3P produced in the ER is DFCP1 (double FYVE-domain-containing protein 1) [17], which is found only in mammals and is present on so-called omegasomes as described below (also see Chapter 2).

Additional protein complexes required for autophagy include two ubiquitin-like conjugation systems that are conserved between species. These systems share one Atg protein (Atg7) and are also inter-dependent. The first, as described in Chapter 4, is the Atg12–Atg5–Atg16 complex, in which the Atg12 molecule is the ubiquitin-like protein. Atg5 is covalently linked to

Atg12 while Atg16 associates with the Atg12–Atg5 conjugate to form a large complex. The second ubiquitin-like conjugation system involves Atg8 that is covalently modified by the lipid PE (phosphatidylethanolamine). Mammalian cells have several Atg8-like molecules, called LC3 (light-chain 3), GABARAP (γ-aminobutyric acid receptor-associated protein) and GATE-16 (Golgi-associated ATPase enhancer of 16 kDa), of which the first two have multiple family members (see Chapter 5). Importantly, Atg12–Atg5–Atg16 provides an E3-like activity to promote the lipidation of LC3 by Atg3 [18]. The best studied Atg8 protein in mammalian cells is LC3B, which when lipidated associates with autophagosomal membranes, and the GFP-tagged protein (GFP–LC3B) has been widely used as a marker for these membranes. Lastly, Atg9 is a transmembrane protein that spans the membrane six times, and has its N- and C-terminal domains in the cytosol. Importantly, it is the only transmembrane Atg protein and as such understanding its localization should provide important information about the membranes contributing to autophagy.

The function of Atg proteins: hierarchal analysis

There are two hierarchies to consider: the one that transmits the induction signal, and the other that occurs on the membrane. Using biochemical approaches, and in particular studying signature phosphorylation events, which report activating phosphorylations, it has been shown that the ULK and Beclin1 complexes are the most upstream of the Atg proteins, directly downstream of nutrient and energy sensors (Figure 2). These kinases (through phosphorylation events) alter the activities of their substrates and directly transmit the signal; however, the substrates of the ULK1/2 protein kinase complex remain uncharacterized. The ubiquitination-like modifications also occur in a biochemically defined sequence, and could also be considered non-enzymatic transmitters of the signal. Atg7 acts as an E1 in both the upstream Atg12–Atg5 conjugate and the LC3–PE conjugation. The transmission of the signal is now described elegantly by the structural analysis demonstrating Atg12 binds Atg3, the E2 of the second conjugation step for LC3–PE formation [19].

The membrane-mediated hierarchal analysis has been informed by the analysis of the recruitment on to the PAS or phagophore, first performed in yeast, and then in mammals [20]. This analysis confirmed the primacy of the ULK complex in the membrane-recruitment hierarchy, followed by the Beclin1 complex. The WIPI proteins and the Atg12–Atg5–Atg16 complex are then recruited downstream of the ULK and Beclin1 complexes. LC3–PE, and the other family members, are generally agreed to be the last Atg proteins to be recruited to the membrane. Intriguingly, the concept of sequential recruitment has been strengthened by recent papers showing that FIP200, a member of the ULK complex, interacts with Atg16 [21,22]. In addition, ULK1 has also been shown to have an LIR (LC3-interacting region) motif and bind LC3 family members [23,24], suggesting in fact that there may be a reinforcement cycle occurring between the Atg proteins that are sequentially recruited.

The notable exception in the hierarchy is Atg9. In yeast, Atg9 has been shown to be required for the assembly of the PAS, catalysing the delivery of vesicles from the Atg9 compartment [25,26] and is required for expansion of the forming and closing autophagosome. In mammals, in fed cells, Atg9 traffics between the Golgi and endosomes, and in starved cells additionally from the equivalent Atg9 compartment (Figure 1) [27]. An important difference between the yeast and mammalian pathway is the dynamics of Atg9 from the expanding autophagosome: in yeast, Atg9 appears to be delivered to the autophagosome membrane [25,26], although in mammals it appears to only participate in kiss-and-run fusion events, and is not incorporated into the expanding phagophore or autophagosome [28].

The origin and source of the phagophore

Live-cell imaging, which can be used to visualize the dynamics of the forming autophagosome during amino-acid starvation, has also suggested that these form simultaneously at multiple sites in cells. This implies that PAS, defined as the primary site or origin of the membranes, may also be widely distributed throughout the cell [5–8]. Nonetheless, although a consensus view is emerging of what phagophores are, many questions remain unanswered regarding the origin and source of the membrane for the PAS and phagophore. One significant issue is the lack of a unique marker to distinguish the PAS from the phagophore, and both from later stages of the pathway. As far as we know, the PAS and the phagophore, and the phagophore and the autophagosome, have overlapping compositions. In addition, the relationship between the phagophore and the omegasome is not yet clear, but is of immediate interest to the field. Live-cell imaging, advanced by immunogold labelling and 3D electron tomography, has shown DFCP1-positive domains (the omegasomes) form from ER membranes which are intimately and perhaps physically linked with phagophores [29,30]. DFCP1 is a unique marker for the omegasome, but is not required for autophagy [17].

Recently, several membranes have been proposed as sites for the nucleation of the phagophore: plasma-membrane-derived vesicles, mitochondria, Golgi and ER (for review see [31]). Importantly, plasma membrane, mitochondria and ER are all distributed widely throughout cells. Regarding the unexpected sources of membrane: the endocytosed plasma membrane, released by fission of plasma membrane domains which is dependent on the coat protein AP-2 and dynamin, has been shown to provide Atg16-positive vesicles which coalesce through a SNARE (soluble N-ethylmaleimide-sensitive fusion protein-attachment protein receptor)-mediated fusion reaction into phagophores [32,33]. Similarly, it was shown that vesicle buds form from mitochondria outer membranes, dependent on mitofusin2, a protein that mediates ER and mitochondrial connections and potentially the production of curvature-initiating lipids. These vesicles are initially Atg5-positive and become LC3-positive [34]. However, in both of these cases, the recruitment of early Atg proteins (ULK complex, Beclin1 complex and WIPI proteins) has not yet been confirmed. The importance of these proteins at early stages has been shown by immunofluorescence and cryo-immunoelectron microscopy in cells expressing a catalytic mutant of Atg4 (which prevents phagophore closure and causes an accumulation of the phagophores by sequestering LC3 and its family members) showing ULK1, WIPI2 and Atg12–Atg5–Atg16 accumulates on phagophores [18,29,35]. Additionally, in mammalian cells, siRNA of ULK1, WIPI2 and its putative interactor Atg2 causes accumulation of Atg9 on phagophores [28,36].

In contrast with plasma membrane and mitochondria, Golgi membranes, implicated in the early cytochemical studies, have been shown in numerous recent studies to be important for autophagosome formation. In particular, these act as a source of Atg9 vesicles and other vesicles whose trafficking is controlled by multi-subunit vesicle-target membrane tethers such as TRAPP (transport protein particle) complexes or the exocyst complex (for details see [31]).

With regard to the ER as a formation site, several laboratories have shown that phagophores emerge from omegasomes, which are subdomains of the ER [17,29,30]. The ULK1/2 and Beclin1 complex members are the first Atg proteins to be recruited to the subdomains on the ER membrane [37,38]. VMP1 (vacuolar membrane protein 1) is an ER-resident transmembrane protein, and again although not an Atg protein, is required for autophagy under particular conditions, and is recruited to these early phagophores on the ER [37]. In addition, these ER subdomains, which contain the Atg proteins listed above, plus DFCP1 and VMP1, have been shown recently to contain MAM (mitochondria-associated ER membrane) proteins, found in close contact with mitochondria, and regulated by the SNARE protein syntaxin 17 [39] (Figure 3). These data have combined to present a strong case for the ER, or subdomains of the ER, being required for formation of the phagophore, at least during amino-acid starvation. More research remains to be carried out in order to understand how the phagophores can form, and

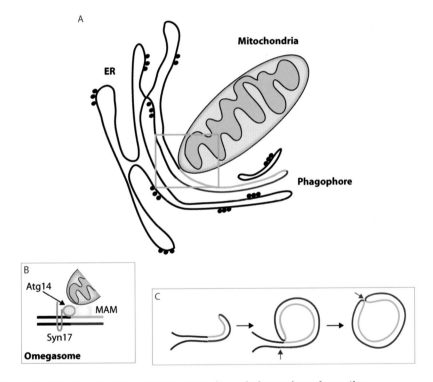

Figure 3. ER–mitochondria–omegasome sites of phagophore formation
(**A**) The current model for the formation of the phagophore at the omegasome site on the ER. The red and green coloured membranes indicate the surface of the bilayer which will become the outer and inner membrane of the closed autophagosome respectively. Note, in (**A**) and (**B**) this is a hypothetical topology and there are no experimental data to support the association of the MAM and mitochondria only with the presumptive inner autophagosomal membrane. (**B**) Formation in mediated by the SNARE syntaxin 17, Atg14 and the MAM. (**C**) Sites of membrane fission and fusion on the expanding phagophore are indicated by red arrows. To date, no machinery has been described which mediates these events. Of note, the topology of the formation sites is only shown in 2D. When the cell is in 3D, the organization of these domains and membranes is much more complex.

© 2013 Biochemical Society

detach from, ER subdomains, and to elucidate what is the composition of the inner and outer membrane of the autophagosome.

Conclusion

Our understanding of the complex biology underlying the essential process of autophagy is rapidly increasing and becoming ever more supported by molecular details. Our knowledge of both the signals sensing alterations in the growth conditions of cells, and the machinery which responds to the signals have clearly demonstrated that autophagy is a tightly regulated process. The most recent results strongly implicate the ER, or subdomains of the ER, in the formation of the phagophore, but questions remain such as: is the ER the PAS, or do separate PAS structures exist which associates with the expanding ER? Further questions regarding, for example, the mechanism of expansion of the ER-derived phagophore into an autophagosome also requires more study. However, there has been significant progress over the last decades and this progress will eventually be translated into a complete understanding of the pathway.

Summary
- The PAS is the site of the origin of the autophagy pathway.
- Phagophores originate from sites on the ER, in proximity to mitochondria.
- Membrane for the phagophore growth is supplied by the Golgi, plasma membrane and endosomes.

References

1. Clark, Jr, S.L. (1957) Cellular differentiation in the kidneys of newborn mice studies with the electron microscope. J. Cell Biol. **3**, 349–362
2. Novikoff, A.B. (1960) Biochemical and staining reactions of cytoplasmic constituents. In Developing Cell Systems and Their Control (Rudnick, D. and Bonner, J.T., eds), Ronald Press, NY
3. De Duve, C. (1963) The lysosome. Sci. Am. **208**, 64–72
4. Pfeifer, U. (1978) Inhibition by insulin of the formation of autophagic vacuoles in rat liver. A morphometric approach to the kinetics of intracellular degradation by autophagy. J. Cell Biol. **78**, 152–167
5. Mizushima, N., Yamamoto, A., Hatano, M., Kobayashi, Y., Kabeya, Y., Suzuki, K., Tokuhisa, T., Ohsumi, Y. and Yoshimori, T. (2001) Dissection of autophagosome formation using Apg5-deficient mouse embryonic stem cells. J. Cell Biol. **152**, 657–668
6. Köchl, R., Hu, X., Chan, E. and Tooze, S.A. (2006) Microtubules facilitate autophagosomal formation and fusion of autophagosomes with endosomes. Traffic **7**, 129–145
7. Mizushima, N., Yamamoto, A., Matsui, M., Yoshimori, T. and Ohsumi, Y. (2004) In vivo analysis of autophagy in response to nutrient starvation using transgenic mice expressing a fluorescent autophagosome marker. Mol. Biol. Cell **15**, 1101–1111
8. Fass, E., Shvets, E., Degani, I., Hirschberg, K. and Elazar, Z. (2006) Microtubules support production of starvation-induced autophagosomes but not their targeting and fusion with lysosomes. J. Biol. Chem. **281**, 36303–36316
9. Reunanen, H., Punnonen, E.L. and Hirsimaki, P. (1985) Studies on vinblastine-induced autophagocytosis in mouse liver. V. A cytochemical study on the origin of membranes. Histochemistry **83**, 513–517

10. Arstila, A.U. and Trump, B.F. (1968) Studies on cellular autophagocytosis. Am. J. Pathol. **53**, 687–733
11. Dunn, Jr, W.A. (1990) Studies on the mechanisms of autophagy: formation of the autophagic vacuole. J. Cell Biol. **110**, 1923–1933
12. Yamamoto, A., Masaki, R. and Tashiro, Y. (1990) Characterization of the isolation membranes and the limiting membranes of autophagosomes in rat hepatocytes by lectin cytochemistry. J. Histochem. Cytochem. **38**, 573–580
13. Novikoff, A.B. and Shin, W.Y. (1964) The endoplasmic reticulum in the Golgi zone and its relations to microbodies, Golgi apparatus, and autophagic vacuoles in rat liver cells. J. Microsc. **3**, 187–206
14. Seglen, P.O., Gordon, P.B. and Holen, I. (1990) Non-selective autophagy. Semin. Cell Biol. **1**, 441–448
15. Fengsrud, M., Erichsen, E.S., Berg, T.O., Raiborg, C. and Seglen, P.O. (2000) Ultrastructural characterization of the delimiting membranes of isolated autophagosomes and amphisomes by freeze-fracture electron microscopy. Eur. J. Cell Biol. **79**, 871–882
16. Simonsen, A. and Tooze, S.A. (2009) Coordination of membrane events during autophagy by multiple class III PI3-kinase complexes. J. Cell Biol. **186**, 773–782
17. Axe, E.L., Walker, S.A., Manifava, M., Chandra, P., Roderick, H.L., Habermann, A., Griffiths, G. and Ktistakis, N.T. (2008) Autophagosome formation from membrane compartments enriched in phosphatidylinositol 3-phosphate and dynamically connected to the endoplasmic reticulum. J. Cell Biol. **182**, 685–701
18. Fujita, N., Hayashi-Nishino, M., Fukumoto, H., Omori, H., Yamamoto, A., Noda, T. and Yoshimori, T. (2008) An Atg4B mutant hampers the lipidation of LC3 paralogues and causes defects in autophagosome closure. Mol. Biol. Cell **19**, 4651–4659
19. Sakoh-Nakatogawa, M., Matoba, K., Asai, E., Kirisako, H., Ishii, J., Noda, N.N., Inagaki, F., Nakatogawa, H. and Ohsumi, Y. (2013) Atg12-Atg5 conjugate enhances E2 activity of Atg3 by rearranging its catalytic site. Nat. Struct. Mol. Biol. **20**, 433–439
20. Mizushima, N., Yoshimori, T. and Ohsumi, Y. (2011) The role of Atg proteins in autophagosome formation. Annu. Rev. Cell Dev. Biol. **27**, 107–132
21. Nishimura, T., Kaizuka, T., Cadwell, K., Sahani, M.H., Saitoh, T., Akira, S., Virgin, H.W. and Mizushima, N. (2013) FIP200 regulates targeting of Atg16L1 to the isolation membrane. EMBO Rep. **14**, 284–291
22. Gammoh, N., Florey, O., Overholtzer, M. and Jiang, X. (2013) Interaction between FIP200 and ATG16L1 distinguishes ULK1 complex-dependent and -independent autophagy. Nat. Struct. Mol. Biol. **20**, 144–149
23. Alemu, E.A., Lamark, T., Torgersen, K.M., Birgisdottir, A.B., Larsen, K.B., Jain, A., Olsvik, H., Overvatn, A., Kirkin, V. and Johansen, T. (2012) ATG8 family proteins act as scaffolds for assembly of the ULK complex: sequence requirements for LC3-interacting region (LIR) motifs. J. Biol. Chem. **287**, 39275–39290
24. Kraft, C., Kijanska, M., Kalie, E., Siergiejuk, E., Lee, S.S., Semplicio, G., Stoffel, I., Brezovich, A., Verma, M., Hansmann, I. et al. (2012) Binding of the Atg1/ULK1 kinase to the ubiquitin-like protein Atg8 regulates autophagy. EMBO J. **31**, 3691–3703
25. Mari, M., Griffith, J., Rieter, E., Krishnappa, L., Klionsky, D.J. and Reggiori, F. (2010) An Atg9-containing compartment that functions in the early steps of autophagosome biogenesis. J. Cell Biol. **190**, 1005–1022
26. Yamamoto, H., Kakuta, S., Watanabe, T.M., Kitamura, A., Sekito, T., Kondo-Kakuta, C., Ichikawa, R., Kinjo, M. and Ohsumi, Y. (2012) Atg9 vesicles are an important membrane source during early steps of autophagosome formation. J. Cell Biol. **198**, 219–233
27. Young, A.R.J., Chan, E.Y.W., Hu, X.W., Köchl, R., Crawshaw, S.G., High, S., Hailey, D.W., Lippincott-Schwartz, J. and Tooze, S.A. (2006) Starvation and ULK1-dependent cycling of mammalian Atg9 between the TGN and endosomes. J. Cell Sci. **119**, 3888–3900

28. Orsi, A., Razi, M., Dooley, H., Robinson, D., Weston, A., Collinson, L. and Tooze, S. (2012) Dynamic and transient interactions of Atg9 with autophagosomes, but not membrane integration, is required for autophagy. Mol. Biol. Cell **23**, 1860–1873
29. Hayashi-Nishino, M., Fujita, N., Noda, T., Yamaguchi, A., Yoshimori, T. and Yamamoto, A. (2009) A subdomain of the endoplasmic reticulum forms a cradle for autophagosome formation. Nat. Cell Biol. **11**, 1433–1437
30. Yla-Anttila, P., Vihinen, H., Jokitalo, E. and Eskelinen, E.L. (2009) 3D tomography reveals connections between the phagophore and endoplasmic reticulum. Autophagy **5**, 1180–1185
31. Mari, M., Tooze, S.A. and Reggiori, F. (2011) The puzzling origin of the autophagosomal membrane. F1000 Biol. Rep. **3**, 25
32. Ravikumar, B., Moreau, K., Jahreiss, L., Puri, C. and Rubinsztein, D.C. (2010) Plasma membrane contributes to the formation of pre-autophagosomal structures. Nat. Cell Biol. **12**, 747–757
33. Moreau, K., Ravikumar, B., Renna, M., Puri, C. and Rubinsztein, D.C. (2011) Autophagosome precursor maturation requires homotypic fusion. Cell **146**, 303–317
34. Hailey, D.W., Rambold, A.S., Satpute-Krishnan, P., Mitra, K., Sougrat, R., Kim, P.K. and Lippincott-Schwartz, J. (2010) Mitochondria supply membranes for autophagosome biogenesis during starvation. Cell **141**, 656–667
35. Polson, H.E.J., de Lartigue, J., Rigden, D.J., Reedijk, M., Urbe, S., Clague, M.J. and Tooze, S.A. (2010) Mammalian Atg18 (WIPI2) localizes to omegasome-anchored phagophores and positively regulates LC3 lipidation. Autophagy **6**, 506–522
36. Velikkakath, A.K.G., Nishimura, T., Oita, E., Ishihara, N. and Mizushima, N. (2012) Mammalian Atg2 proteins are essential for autophagosome formation and important for regulation of size and distribution of lipid droplets. Mol. Biol. Cell **23**, 896–909
37. Itakura, E. and Mizushima, N. (2010) Characterization of autophagosome formation site by a hierarchical analysis of mammalian Atg proteins. Autophagy **6**, 764–776
38. Matsunaga, K., Morita, E., Saitoh, T., Akira, S., Ktistakis, N.T., Izumi, T., Noda, T. and Yoshimori, T. (2010) Autophagy requires endoplasmic reticulum targeting of the PI3-kinase complex via Atg14L. J. Cell Biol. **190**, 511–521
39. Hamasaki, M., Furuta, N., Matsuda, A., Nezu, A., Yamamoto, A., Fujita, N., Oomori, H., Noda, T., Haraguchi, T., Hiraoka, Y. et al. (2013) Autophagosomes form at ER–mitochondria contact sites. Nature **495**, 389–393

Two ubiquitin-like conjugation systems that mediate membrane formation during autophagy

Hitoshi Nakatogawa[1]

Frontier Research Center, Tokyo Institute of Technology, Yokohama 226–8503, Japan

Abstract

In autophagy, the autophagosome, a transient organelle specialized for the sequestration and lysosomal or vacuolar transport of cellular constituents, is formed via unique membrane dynamics. This process requires concerted actions of a distinctive set of proteins named Atg (autophagy-related). Atg proteins include two ubiquitin-like proteins, Atg12 and Atg8 [LC3 (light-chain 3) and GABARAP (γ-aminobutyric acid receptor-associated protein) in mammals]. Sequential reactions by the E1 enzyme Atg7 and the E2 enzyme Atg10 conjugate Atg12 to the lysine residue in Atg5, and the resulting Atg12–Atg5 conjugate forms a complex with Atg16. On the other hand, Atg8 is first processed at the C-terminus by Atg4, which is related to ubiquitin-processing/deconjugating enzymes. Atg8 is then activated by Atg7 (shared with Atg12) and, via the E2 enzyme Atg3, finally conjugated to the amino group of the lipid PE (phosphatidylethanolamine). The Atg12–Atg5–Atg16 complex acts as an E3 enzyme for the conjugation reaction of Atg8; it enhances the E2 activity of Atg3 and specifies the site of Atg8–PE production to be autophagy-related membranes. Atg8–PE is suggested to be involved in autophagosome formation at multiple steps, including membrane expansion and closure. Moreover, Atg4 cleaves Atg8–PE to liberate Atg8 from membranes for reuse, and this reaction can also regulate autophagosome formation. Thus these two ubiquitin-like systems are intimately involved in driving the biogenesis of the autophagosomal membrane.

Keywords:
Atg protein, autophagosome formation, conjugation, deconjugation, ubiquitin-like protein.

[1]email hnakatogawa@iri.titech.ac.jp

Introduction

A hallmark of the macroautophagy pathway (hereafter referred to as autophagy) is the biogenesis of the double-membrane vesicle autophagosome that sequesters material to be transported to the lysosome in mammalian cells or the vacuole in yeast and plant cells for degradation [1,2]. Among 36 Atg (autophagy-related) proteins identified in yeast, 15 'core' Atg proteins play pivotal roles in the formation of the autophagosomal membrane and thus are required for any type of autophagy (i.e. both non-selective and selective autophagy). The core Atg proteins constitute six functional units: (i) the Atg1 protein kinase complex, (ii) the Atg9-containing membrane vesicle, (iii) the Atg14-containing PI3K (phosphoinositide 3-kinase) complex, (iv) the Atg2–Atg18 [WIPIs (WD-repeat protein interacting with phosphoinositides) in mammals] complex, (v) the Atg12 conjugation system, and (vi) the Atg8 [LC3 (light-chain 3) and GABARAP (γ-aminobutyric acid receptor-associated protein) in mammals] conjugation system. These units localize to a site for autophagosome formation in an ordered manner to organize the PAS (pre-autophagosomal structure), in which they are likely to produce a precursory membrane, followed by its expansion into the isolation membrane (or phagophore) and the completion of the autophagosome (Figure 1). Therefore analysis of the functions and regulation of the core Atg proteins is essential for understanding the molecular mechanism of autophagosome formation.

It is the intriguing fact that autophagy, a major proteolytic pathway comparable with the ubiquitin–proteasome pathway, employs two ubiquitin-like systems, the Atg12 and Atg8 systems, to which approximately half of the core Atg proteins are devoted. These systems have been extensively studied from different viewpoints, including those of cell biology, biochemistry and structural biology. Various unique and elaborate mechanisms underlying these systems have been unveiled, which attract researchers in not only autophagy, but also other fields. In the present chapter, I summarize our current knowledge on the functions and regulation of these ubiquitin-like systems in autophagosome formation and their mechanisms. Most of the important findings were first achieved by yeast studies and then confirmed in mammals; essentially, the same mechanisms are working. Therefore I describe the following mainly on the basis of the results of the yeast system, except for the case where there is a remarkable difference from the mammalian system.

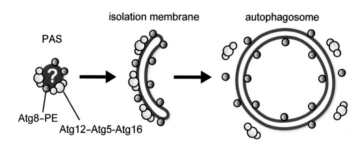

Figure 1. Process of autophagosome formation and localization of Atg12–Atg5–Atg16 and Atg8–PE

Both the Atg12–Atg5–Atg16 complex and the Atg8–PE conjugate localize to the PAS. Although Atg8–PE is comparatively evenly distributed on the isolation membrane and retained on the complete autophagosome, Atg12–Atg5–Atg16 preferentially associates with the convex surface of the isolation membrane and dissociates from the membrane upon autophagosome completion.

Conjugation reactions of Atg12 and Atg8

Atg12 and Atg8 are conjugated to each specific target via a series of enzymatic reactions similar to protein ubiquitination (Figure 2). The C-terminal carboxyl group of Atg12 is activated by the E1 enzyme Atg7 with consumption of ATP to form a thioester bond with its catalytic cysteine residue, then transferred to the catalytic cysteine residue of the E2 enzyme Atg10, and eventually attached to the amino group of the lysine residue in Atg5 via an isopeptide bond [3]. Atg8 is synthesized with an additional sequence at its C-terminus, which is immediately removed by the processing/deconjugating enzyme Atg4 to expose the glycine residue essential for subsequent reactions [4]. The conjugation reaction of Atg8 is catalysed by Atg7 (Atg12 and Atg8 share the same E1 enzyme) and the specific E2 Atg3 (Figure 2) [5]. Remarkably, the target of Atg8 is not a protein, but the lipid PE (phosphatidylethanolamine); the C-terminal carboxyl group of Atg8 forms an amide bond with the amino group in the hydrophilic head moiety of PE, thereby Atg8 is anchored to membranes.

Structural analyses of Atg7 revealed that it has an adenylation domain similar to other E1 enzymes and forms a homodimer via this domain [6,7]. In addition, Atg7 has a unique domain

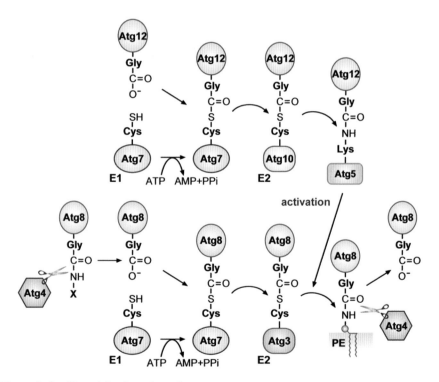

Figure 2. Atg12 and Atg8 conjugation systems
Similar to protein ubiquitination, the C-terminal carboxyl group of the ubiquitin-like protein Atg12 is activated by the E1 enzyme Atg7 using ATP, forms a thioester intermediate with its catalytic cysteine residue, is transferred to the catalytic cysteine residue of the E2 enzyme Atg10, and finally forms an isopeptide bond with the lysine residue in Atg5. Likewise, the ubiquitin-like protein Atg8 is activated by Atg7, transferred to the E2 enzyme Atg3, and finally conjugated to PE via an amide bond. Atg4 mediates both the C-terminal processing and deconjugation of Atg8. The Atg12–Atg5 conjugate enhances the E2 activity of Atg3.

in its N-terminal region, to which Atg10 and Atg3 bind in a mutually exclusive manner [6–9]. In combination with biochemical analyses, it was proposed that the Atg12 or Atg8 that has formed a thioester intermediate with a protomer in the Atg7 dimer is transferred to Atg10 or Atg3 bound to the other protomer in the same dimer respectively.

In typical E2 enzymes for ubiquitin, an Asn* (invariable asparagine residue) plays a pivotal role in the transfer of ubiquitin forming a thioester bond with the catalytic cysteine residue to a substrate lysine residue. However, in Atg3, a threonine residue, which is highly conserved among Atg3 homologues, is located at a position corresponding to Asn* [10]. Biochemical analysis suggested that this threonine residue is involved in Atg8 transfer to PE in a manner similar to Asn* in other E2 enzymes [11]. Atg3 may require the threonine residue to target not lysine residues in proteins, but the lipid PE.

Whereas E3 enzymes in the ubiquitin system determine substrate specificity and stimulate ubiquitin transfer to substrates, there is no E3 enzyme for the Atg12 system. Instead, the E2 enzyme Atg10 directly recognizes the substrate Atg5 [12]. This is reasonable considering that Atg5 is the sole substrate for the Atg12 conjugation reaction. Similarly, the E2 enzyme Atg3 itself recognizes the substrate PE in the Atg8 system, although *in vitro* reactions under non-physiological conditions allow Atg3 to conjugate Atg8 to phosphatidylserine, which has an amino group in its hydrophilic head as well as PE [13,14]. Although the Atg12–Atg5 conjugate enhances the E2 activity of Atg3 (see below), it does not affect the substrate preference of Atg3 [15].

Functions of Atg12–Atg5 and Atg8–PE conjugates

Cells defective in the formation of the Atg12–Atg5 conjugate (i.e. those lacking Atg12, Atg5 or Atg10) also exhibit a defect in the formation of Atg8–PE (called LC3/GABARAP-II in mammals) [16,17]. This indicated a linkage between the two conjugation systems. The conjugation reaction of Atg8 can be reconstituted *in vitro* using purified proteins, Atg8, Atg7 and Atg3, PE-containing liposomes (artificial membrane vesicles) and ATP [18]. On the other hand, the simultaneous expression of Atg12, Atg5, Atg7 and Atg10 allows the formation of Atg12–Atg5 in *Escherichia coli* cells, from which the conjugate can be purified. By adding it to the *in vitro* Atg8 system, it was clearly shown that Atg12–Atg5 itself has an ability to accelerate Atg8–PE formation [15]. It was further shown that Atg12–Atg5 interacts with the E2 enzyme Atg3 and stimulates the transfer of Atg8 forming a thioester bond with the catalytic cysteine residue of Atg3 to PE (Figure 2). Thus in the Atg8 conjugation reaction, Atg12–Atg5 exerts an activity like E3 enzymes in the ubiquitin system. *In vivo*, Atg12–Atg5 functions as a complex with Atg16 (Atg16L in mammals), which forms a dimer in its C-terminal region [19–21]. Since Atg16 interacts with Atg5 in the N-terminal region, the Atg12–Atg5–Atg16 complex is a 2:2:2 heterohexamer. Atg16 is dispensable for the E3-like activity of Atg12–Atg5, but is essential for the localization of the complex to autophagy-related membranes (see below).

Fluorescence and immunoelectron microscopy showed that Atg8–PE localizes to all the autophagy-related structures: the PAS, the isolation membrane, the complete autophagosome, the autophagic body (the inner membrane vesicle released into the vacuole on fusion between the autophagosomal outer membrane and the vacuole in yeast and plant cells) and the autolysosome (the lysosome fused with the autophagosome in mammalian cells) [22,23] (Figure 1). *In vitro* studies also provided a clue to elucidate the molecular function of the Atg8–PE conjugate during autophagosome formation [24]. In the *in vitro* reaction, Atg8 is conjugated to PE in the outer leaflet of the liposomal lipid bilayer (Figure 3A). It was found that this leads to aggregation of the liposomes, associated with hemifusion of the membranes (fusion between the outer leaflets of two opposed membranes with the inner leaflets left intact) (Figure 3A). It was also shown that Atg8 forms oligomers when conjugated to PE. These results suggest that Atg8–PE conjugates on one membrane can interact with those on another membrane to tether the membranes together and cause their hemifusion. Most Atg8 mutants defective in autophagosome formation exhibited a significant decrease in membrane tethering and hemifusion, suggesting that these functions of Atg8 observed *in vitro* are related to its *in vivo* role during autophagosome formation [24].

How are the membrane tethering and hemifusion functions of Atg8–PE involved in autophagosome formation? Autophagosomes smaller than those in wild-type cells are formed in cells expressing Atg8 mutants partially defective in membrane tethering and hemifusion, indicating that the Atg8–PE functions are involved in the step that determines the size of the autophagosome, i.e. the step of membrane expansion (Figure 3B) [24]. Consistently, decreased expression levels of Atg8 result in the formation of small autophagosomes [25]. It is also possible that Atg8–PE plays an important role in an earlier event, the formation of an as yet unidentified precursory membrane at the PAS. In addition, it was reported that unclosed isolation membranes with abnormal morphology accumulate in mammalian cells deficient for PE conjugation of Atg8 homologues, suggesting that Atg8–PE is also involved in the closure of the isolation membrane (this step requires membrane fission) in addition to its normal

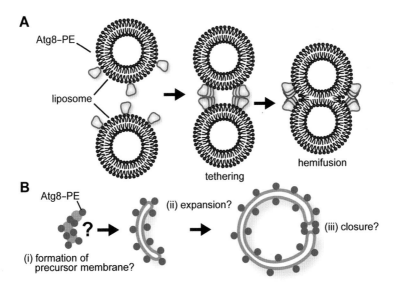

Figure 3. Roles of Atg8–PE in autophagosome formation
(**A**) Atg8–PE oligomerizes and causes tethering and hemifusion of liposomal membranes *in vitro*. (**B**) Possible roles of Atg8–PE during autophagosome formation (see text for details).

© 2013 Biochemical Society

development [26,27]. Further analyses are still required to understand precisely how Atg8–PE is involved in autophagosome formation.

It should be noted that Atg8–PE-mediated hemifusion, compared with liposome aggregation, is sensitive to the lipid composition of liposomes (H. Nakatogawa and Y. Ohsumi, unpublished results). It was also reported that hemifusion does not occur in liposomes containing PE at concentrations in typical organelle membranes [28]. Although this may indicate that Atg8–PE does not cause hemifusion *in vivo*, to clarify this point, it is essential to determine the lipid composition of membranes where Atg8–PE functions. At any rate, only hemifusion does not lead to the expansion of membranes; another protein(s) and/or a specific lipid composition may help complete fusion to occur in cells. It was recently reported that SNARE (soluble *N*-ethylmaleimide-sensitive fusion protein-attachment protein receptor) proteins, which mediate membrane fusion in various endomembrane systems, are required for autophagosome formation, but it is unclear whether they co-operate with Atg8 [28,29].

Recently, on the basis of theoretical analysis, an intriguing model for curving of the isolation membrane was proposed [30]. The isolation membrane can spontaneously curve when it expands to a critical size, which is determined by three properties: the lateral dimension of the membrane, the molecular composition of the highly curved rims and an asymmetry between the two flat faces. Proteins that bind to membranes such as Atg8–PE can regulate the latter two properties of the sheet. Whereas a higher protein concentration at the rims suppresses curving of the isolation membrane and thus results in formation of a larger autophagosome, an asymmetric protein distribution between the two flat faces facilitates it, leading to small autophagosome formation. Therefore in these, Atg8–PE density and localization on the isolation membrane can contribute to determining the size of the autophagosome. These possible roles of Atg8–PE are not mutually exclusive with those discussed above, and it is tempting to speculate that Atg8–PE is involved in autophagosome formation in multiple ways.

Mechanism of Atg3 activation by Atg12–Atg5

Although the primary sequences of Atg8 and Atg12 show little similarity to ubiquitin, structural studies revealed that these proteins actually adopt a ubiquitin-like fold [10]. In addition, it was also revealed that Atg5 contains two ubiquitin-like folds [10,31,32]. These facts highlighted the peculiarity of the Atg conjugation systems; a ubiquitin-like protein conjugate with three ubiquitin-like folds serves as an E3 enzyme in the conjugation reaction of another ubiquitin-like protein. We recently succeeded in unveiling how Atg12–Atg5 enhances the E2 activity of Atg3 [11]. Atg3 has an E2 core domain similar to other E2 enzymes [10]. However, Atg3 adopts an inactive conformation in the absence of Atg12–Atg5; the side chain of the catalytic cysteine residue of Atg3 faces in the opposite direction from the aforementioned threonine residue, whereas that of typical E2 enzymes is directed towards Asn* (Figure 4A). Atg12–Atg5 causes the reorientation of the cysteine residue toward the threonine residue, resulting in the enhancement of Atg3 activity [11] (Figure 4B). How are three ubiquitin-like folds in Atg12–Atg5 involved in the rearrangement of the Atg3 catalytic centre? This is the intriguing question to be addressed next.

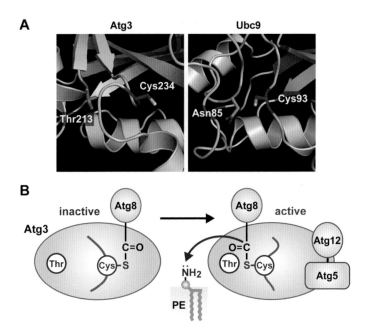

Figure 4. Mechanism of Atg3 activation by Atg12–Atg5
(**A**) Comparison of the catalytic centres of *Saccharomyces cerevisiae* Atg3 and mammalian Ubc9 (SUMO E2 enzyme). Atg3 uses a threonine residue (Thr^{213}), instead of the asparagine residue (Asn^{85} in Ubc9) that is essential for ubiquitin/ubiquitin-like protein transfer to a substrate lysine residue, to transfer Atg8 to PE. Whereas the catalytic cysteine residue (Cys^{93}) of Ubc9 is oriented towards Asn^{85}, that of Atg3 (Cys^{234}) faces in the opposite direction from Thr^{213}, suggesting that this structure represents an inactive form of Atg3. (**B**) Atg12–Atg5 induces a conformational change in the Atg3 catalytic centre, thereby Cys^{234} is reoriented towards Thr^{213}, which facilitates Atg8 transfer to PE.

Spatial regulation of Atg8–PE formation by the Atg12–Atg5–Atg16 complex

In response to autophagy-inducing signals, the expression of Atg8 is transcriptionally up-regulated [22]. The Ume6–Sin3–Rpd3 complex binds to the promoter region of the *ATG8* gene to repress its transcription under nutrient-rich conditions [33]. Rim15 phosphorylates Ume6 to disassemble the complex under starvation conditions, resulting in up-regulation of *ATG8* transcription. Autophagy-inducing signals also stimulate the formation of Atg8–PE [24]. This is tightly linked to the localization of the Atg12–Atg5–Atg16 complex, which changes its localization from the cytoplasm to the PAS and the isolation membrane following autophagy-inducing signals [16,17] (Figure 1). It was also reported that Atg12–Atg5–Atg16 predominantly localizes to the convex surface of curved isolation membranes in mammalian cells [17]. Targeting Atg12–Atg5–Atg16 to those structures would lead to the production of Atg8–PE on the membranes via its E3 enzyme-like function that activates Atg3. Since PE is a major component in most intracellular membranes, the localized activation of Atg3 by Atg12–Atg5–Atg16 is likely to be important to produce Atg8–PE on autophagy-related membranes.

Atg5 and Atg16 co-operatively act to target the complex to the PAS [34]. *In vitro* analysis showed that Atg5 alone and the Atg12–Atg5–Atg16 complex, but neither Atg16 alone nor Atg12–Atg5 can bind to liposomes, suggesting that Atg16 increases the membrane-binding ability of Atg5, which is suppressed by Atg12 in the conjugate [35]. It is also known that the production of PI3P (phosphatidylinositol 3-phosphate) by the Atg14-containing PI3K complex is a prerequisite for the localization of Atg12–Atg5–Atg16 [34,36]. However, how Atg12–Atg5–Atg16 is localized to the PAS and how it is associated with the isolation membrane remain to be elucidated. Unlike Atg8, Atg12–Atg5–Atg16 is released from the membrane immediately before or on completion of the autophagosome [17]. This mechanism is also still elusive, but may involve PI3Ps [37].

Significance of Atg8 deconjugation by Atg4

As described above, Atg4 cleaves Atg8 at the peptide bond C-terminal to the glycine residue essential for the conjugation reaction. In addition, Atg4 also serves as a deconjugating enzyme that cleaves the amide bond between Atg8 and PE to release the protein from PE in membranes [4] (Figure 2). This reaction is thought to be important for reusing the Atg8 that has exerted its function for autophagosome formation and would occur on the complete autophagosome and the autolysosome or vacuole (Atg8–PE on the autophagosomal outer membrane can be transferred on to the lysosomal or vacuolar membrane following their fusion). Moreover, Atg8 deconjugation by Atg4 may also occur on the isolation membrane, which could positively or negatively affect membrane formation [38–40]. On the other hand, a mechanism by which the Atg8–PE that has not fulfilled its role is protected from deconjugation by Atg4 may also exist. Thus controlling Atg4-mediated deconjugation of Atg8 can regulate autophagosome formation, but further analyses are required to assess these possibilities.

As discussed above, the localized production of Atg8–PE on autophagy-related membranes should be achieved by the localization of the Atg12–Atg5–Atg16 complex to those membranes. However, recent studies suggested that this mechanism is not that strict: conjugation enzymes in the cytoplasm erroneously produce Atg8–PE on various intracellular membranes to a considerable degree [39–41]. Thus another new role of Atg4 was proposed; Atg4 deconjugates those non-productive Atg8–PE to maintain a cytoplasmic reservoir of unconjugated Atg8, which is required for Atg8–PE formation at correct sites (Figure 5).

Conclusions

In the present chapter, I have described the autophagy-related ubiquitin-like systems in the context of autophagosomal membrane biogenesis, but recent studies have also revealed their other aspects. The most prominent is a role of Atg8 homologues in selective types of autophagy, in which degradation targets, such as ubiquitin-positive protein aggregates, damaged mitochondria, superfluous peroxisomes and invasive bacterial cells, are exclusively enwrapped by the autophagosomal membrane [42,43]. Atg8 homologues bind to a consensus sequence named the AIM (Atg8 family-interacting motif) or LIR (LC3-interacting region) in receptor proteins, which specifically recognize each target, with highly conserved binding

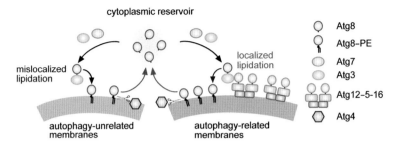

Figure 5. Roles of Atg8 deconjugation by Atg4
The localization of the Atg12–Atg5–Atg16 complex results in preferential Atg8–PE formation on autophagy-related membranes. Atg4 cleaves Atg8–PE on those membranes to recycle Atg8 and regulate membrane formation. On the other hand, conjugation enzymes also produce Atg8–PE on various endomembranes independent of autophagy, which is recycled by Atg4 to maintain the cytoplasmic pool of Atg8 available for autophagosome formation at correct sites.

pockets [44]. The interaction of Atg8–PE on the growing autophagosomal membrane with receptor proteins is likely to link the membrane to the targets. As another aspect, although there is a single Atg8 in yeast cells, many other organisms, including mammals and plants, have multiple Atg8 homologues. Although it had been suggested that they have different functions, actual cases have been reported recently [45,46]. It has also been reported that some viruses and bacteria target the Atg conjugation systems to interfere with autophagy for their proliferation in host cells. The roles of the Atg8 and Atg12 systems in diverse biological events, such as membrane traffic, the regulation of vacuolar morphology, phagocytosis and apoptosis, have also been implicated, which were suggested to be independent of autophagy. Studies on the Atg12 and Atg8 systems will continue to provide important insights into mechanisms, physiology and pathology not only within, but also beyond, autophagy.

Summary

- Autophagosome formation requires the Atg12 and Atg8 ubiquitin-like systems.
- The Atg12 and Atg8 conjugation reactions share Atg7 as an E1 enzyme and employ Atg10 and Atg3 as specific E2 enzymes.
- Atg4 serves as a processing/deconjugating enzyme for Atg8.
- Atg12 and Atg8 form conjugates with Atg5 and PE respectively.
- Atg12–Atg5 induces the rearrangement of the Atg3 catalytic centre to enhance its E2 activity.
- Atg16 forms a complex with Atg12–Atg5 and targets the complex to autophagy-related membranes, where Atg8–PE is produced.
- Atg8–PE is involved in autophagosome formation at multiple steps.
- Atg4-mediated deconjugation recycles the Atg8 that has fulfilled its role in membrane formation or has erroneously formed a conjugate with PE in autophagy-unrelated membranes.
- Thus the Atg12 and Atg8 systems are intimately involved in driving membrane formation during autophagy.

I thank Dr Yoshinori Ohsumi for critically reading this chapter. I apologize that many important references have been omitted due to the limitation on the number of citations.

References

1. Nakatogawa, H., Suzuki, K., Kamada, Y. and Ohsumi, Y. (2009) Dynamics and diversity in autophagy mechanisms: lessons from yeast. Nat. Rev. Mol. Cell Biol. **10**, 458–467
2. Mizushima, N., Yoshimori, T. and Ohsumi, Y. (2011) The role of Atg proteins in autophagosome formation. Annu. Rev. Cell Dev. Biol. **27**, 107–132
3. Mizushima, N., Noda, T., Yoshimori, T., Tanaka, Y., Ishii, T., George, M.D., Klionsky, D.J., Ohsumi, M. and Ohsumi, Y. (1998) A protein conjugation system essential for autophagy. Nature **395**, 395–398
4. Kirisako, T., Ichimura, Y., Okada, H., Kabeya, Y., Mizushima, N., Yoshimori, T., Ohsumi, M., Takao, T., Noda, T. and Ohsumi, Y. (2000) The reversible modification regulates the membrane-binding state of Apg8/Aut7 essential for autophagy and the cytoplasm to vacuole targeting pathway. J. Cell Biol. **151**, 263–276
5. Ichimura, Y., Kirisako, T., Takao, T., Satomi, Y., Shimonishi, Y., Ishihara, N., Mizushima, N., Tanida, I., Kominami, E., Ohsumi, M. et al. (2000) A ubiquitin-like system mediates protein lipidation. Nature **408**, 488–492
6. Noda, N.N., Satoo, K., Fujioka, Y., Kumeta, H., Ogura, K., Nakatogawa, H., Ohsumi, Y. and Inagaki, F. (2011) Structural basis of Atg8 activation by a homodimeric E1, Atg7. Mol. Cell **44**, 462–475
7. Taherbhoy, A.M., Tait, S.W., Kaiser, S.E., Williams, A.H., Deng, A., Nourse, A., Hammel, M., Kurinov, I., Rock, C.O., Green, D.R. and Schulman, B.A. (2011) Atg8 transfer from Atg7 to Atg3: a distinctive E1-E2 architecture and mechanism in the autophagy pathway. Mol. Cell **44**, 451–461
8. Yamaguchi, M., Matoba, K., Sawada, R., Fujioka, Y., Nakatogawa, H., Yamamoto, H., Kobashigawa, Y., Hoshida, H., Akada, R., Ohsumi, Y. et al. (2012) Noncanonical recognition and UBL loading of distinct E2s by autophagy-essential Atg7. Nat. Struct. Mol. Biol. **19**, 1250–1256
9. Kaiser, S.E., Mao, K., Taherbhoy, A.M., Yu, S., Olszewski, J.L., Duda, D.M., Kurinov, I., Deng, A., Fenn, T.D., Klionsky, D.J. and Schulman, B.A. (2012) Noncanonical E2 recruitment by the autophagy E1 revealed by Atg7–Atg3 and Atg7–Atg10 structures. Nat. Struct. Mol. Biol. **19**, 1242–1249
10. Noda, N.N., Ohsumi, Y. and Inagaki, F. (2009) ATG systems from the protein structural point of view. Chem. Rev. **109**, 1587–1598
11. Sakoh-Nakatogawa, M., Matoba, K., Asai, E., Kirisako, H., Ishii, J., Noda, N.N., Inagaki, F., Nakatogawa, H. and Ohsumi, Y. (2013) Atg12–Atg5 conjugate enhances E2 activity of Atg3 by rearranging its catalytic site. Nat. Struct. Mol. Biol. **20**, 433–439
12. Yamaguchi, M., Noda, N.N., Yamamoto, H., Shima, T., Kumeta, H., Kobashigawa, Y., Akada, R., Ohsumi, Y. and Inagaki, F. (2012) Structural insights into Atg10-mediated formation of the autophagy-essential Atg12–Atg5 conjugate. Structure **20**, 1244–1254
13. Sou, Y.S., Tanida, I., Komatsu, M., Ueno, T. and Kominami, E. (2006) Phosphatidylserine in addition to phosphatidylethanolamine is an in vitro target of the mammalian Atg8 modifiers, LC3, GABARAP, and GATE-16. J. Biol. Chem. **281**, 3017–3024
14. Oh-oka, K., Nakatogawa, H. and Ohsumi, Y. (2008) Physiological pH and acidic phospholipids contribute to substrate specificity in lipidation of Atg8. J. Biol. Chem. **283**, 21847–2152
15. Hanada, T., Noda, N.N., Satomi, Y., Ichimura, Y., Fujioka, Y., Takao, T., Inagaki, F. and Ohsumi, Y. (2007) The Atg12–Atg5 conjugate has a novel E3-like activity for protein lipidation in autophagy. J. Biol. Chem. **282**, 37298–37302

16. Suzuki, K., Kirisako, T., Kamada, Y., Mizushima, N., Noda, T. and Ohsumi, Y. (2001) The pre-autophagosomal structure organized by concerted functions of APG genes is essential for autophagosome formation. EMBO J. **20**, 5971–5981
17. Mizushima, N., Yamamoto, A., Hatano, M., Kobayashi, Y., Kabeya, Y., Suzuki, K., Tokuhisa, T., Ohsumi, Y. and Yoshimori, T. (2001) Dissection of autophagosome formation using Apg5-deficient mouse embryonic stem cells. J. Cell Biol. **152**, 657–668
18. Ichimura, Y., Imamura, Y., Emoto, K., Umeda, M., Noda, T. and Ohsumi, Y. (2004) *In vivo* and *in vitro* reconstitution of Atg8 conjugation essential for autophagy. J. Biol. Chem. **279**, 40584–40592
19. Mizushima, N., Noda, T. and Ohsumi, Y. (1999) Apg16p is required for the function of the Apg12p-Apg5p conjugate in the yeast autophagy pathway. EMBO J. **18**, 3888–3896
20. Fujita, N., Saitoh, T., Kageyama, S., Akira, S., Noda, T. and Yoshimori, T. (2009) Differential involvement of Atg16L1 in Crohn disease and canonical autophagy: analysis of the organization of the Atg16L1 complex in fibroblasts. J. Biol. Chem. **284**, 32602–32609
21. Fujioka, Y., Noda, N.N., Nakatogawa, H., Ohsumi, Y. and Inagaki, F. (2010) Dimeric coiled-coil structure of *Saccharomyces cerevisiae* Atg16 and its functional significance in autophagy. J. Biol. Chem. **285**, 1508–1515
22. Kirisako, T., Baba, M., Ishihara, N., Miyazawa, K., Ohsumi, M., Yoshimori, T., Noda, T. and Ohsumi, Y. (1999) Formation process of autophagosome is traced with Apg8/Aut7p in yeast. J. Cell Biol. **147**, 435–446
23. Kabeya, Y., Mizushima, N., Ueno, T., Yamamoto, A., Kirisako, T., Noda, T., Kominami, E., Ohsumi, Y. and Yoshimori, T. (2000) LC3, a mammalian homologue of yeast Apg8p, is localized in autophagosome membranes after processing. EMBO J. **19**, 5720–5728
24. Nakatogawa, H., Ichimura, Y. and Ohsumi, Y. (2007) Atg8, a ubiquitin-like protein required for autophagosome formation, mediates membrane tethering and hemifusion. Cell **130**, 165–178
25. Xie, Z., Nair, U. and Klionsky, D.J. (2008) Atg8 controls phagophore expansion during autophagosome formation. Mol. Biol. Cell **19**, 3290–3298
26. Sou, Y.S., Waguri, S., Iwata, J., Ueno, T., Fujimura, T., Hara, T., Sawada, N., Yamada, A., Mizushima, N., Uchiyama, Y. et al. (2008) The Atg8 conjugation system is indispensable for proper development of autophagic isolation membranes in mice. Mol. Biol. Cell **19**, 4762–4775
27. Fujita, N., Hayashi-Nishino, M., Fukumoto, H., Omori, H., Yamamoto, A., Noda, T. and Yoshimori, T. (2008) An Atg4B mutant hampers the lipidation of LC3 paralogues and causes defects in autophagosome closure. Mol. Biol. Cell **19**, 4651–4659
28. Nair, U., Jotwani, A., Geng, J., Gammoh, N., Richerson, D., Yen, W.L., Griffith, J., Nag, S., Wang, K., Moss, T. et al. (2011) SNARE proteins are required for macroautophagy. Cell **146**, 290–302
29. Moreau, K., Ravikumar, B., Renna, M., Puri, C. and Rubinsztein, D.C. (2011) Autophagosome precursor maturation requires homotypic fusion. Cell **146**, 303–317
30. Knorr, R.L., Dimova, R. and Lipowsky, R. (2012) Curvature of double-membrane organelles generated by changes in membrane size and composition. PLoS ONE **7**, e32753
31. Otomo, C., Metlagel, Z., Takaesu, G. and Otomo, T. (2012) Structure of the human ATG12–ATG5 conjugate required for LC3 lipidation in autophagy. Nat. Struct. Mol. Biol. **20**, 59–66
32. Noda, N.N., Fujioka, Y., Hanada, T., Ohsumi, Y. and Inagaki, F. (2012) Structure of the Atg12–Atg5 conjugate reveals a platform for stimulating Atg8–PE conjugation. EMBO Rep. **14**, 206–211
33. Bartholomew, C.R., Suzuki, T., Du, Z., Backues, S.K., Jin, M., Lynch-Day, M.A., Umekawa, M., Kamath, A., Zhao, M., Xie, Z. et al. (2012) Ume6 transcription factor is part of a signaling cascade that regulates autophagy. Proc. Natl. Acad. Sci. U.S.A. **109**, 11206–11210
34. Suzuki, K., Kubota, Y., Sekito, T. and Ohsumi, Y. (2007) Hierarchy of Atg proteins in pre-autophagosomal structure organization. Genes Cells **12**, 209–218

35. Romanov, J., Walczak, M., Ibiricu, I., Schuchner, S., Ogris, E., Kraft, C. and Martens, S. (2012) Mechanism and functions of membrane binding by the Atg5–Atg12/Atg16 complex during autophagosome formation. EMBO J. **31**, 4304–4317
36. Itakura, E. and Mizushima, N. (2010) Characterization of autophagosome formation site by a hierarchical analysis of mammalian Atg proteins. Autophagy **6**, 764–776
37. Cebollero, E., van der Vaart, A., Zhao, M., Rieter, E., Klionsky, D.J., Helms, J.B. and Reggiori, F. (2012) Phosphatidylinositol-3-phosphate clearance plays a key role in autophagosome completion. Curr. Biol. **22**, 1545–1553
38. Scherz-Shouval, R., Shvets, E., Fass, E., Shorer, H., Gil, L. and Elazar, Z. (2007) Reactive oxygen species are essential for autophagy and specifically regulate the activity of Atg4. EMBO J. **26**, 1749–1760
39. Yu, Z.Q., Ni, T., Hong, B., Wang, H.Y., Jiang, F.J., Zou, S., Chen, Y., Zheng, X.L., Klionsky, D.J., Liang, Y. and Xie, Z. (2012) Dual roles of Atg8-PE deconjugation by Atg4 in autophagy. Autophagy **8**, 883–892
40. Nair, U., Yen, W.L., Mari, M., Cao, Y., Xie, Z., Baba, M., Reggiori, F. and Klionsky, D.J. (2012) A role for Atg8-PE deconjugation in autophagosome biogenesis. Autophagy **8**, 780–793
41. Nakatogawa, H., Ishii, J., Asai, E. and Ohsumi, Y. (2012) Atg4 recycles inappropriately lipidated Atg8 to promote autophagosome biogenesis. Autophagy **8**, 177–186
42. Johansen, T. and Lamark, T. (2011) Selective autophagy mediated by autophagic adapter proteins. Autophagy **7**, 279–296
43. Weidberg, H., Shvets, E. and Elazar, Z. (2011) Biogenesis and cargo selectivity of autophagosomes. Annu. Rev. Biochem. **80**, 125–156
44. Noda, N.N., Ohsumi, Y. and Inagaki, F. (2010) Atg8-family interacting motif crucial for selective autophagy. FEBS Lett. **584**, 1379–1385
45. Weidberg, H., Shvets, E., Shpilka, T., Shimron, F., Shinder, V. and Elazar, Z. (2010) LC3 and GATE-16/GABARAP subfamilies are both essential yet act differently in autophagosome biogenesis. EMBO J. **29**, 1792–1802
46. von Muhlinen, N., Akutsu, M., Ravenhill, B.J., Foeglein, A., Bloor, S., Rutherford, T.J., Freund, S.M., Komander, D. and Randow, F. (2012) LC3C, bound selectively by a noncanonical LIR motif in NDP52, is required for antibacterial autophagy. Mol. Cell **48**, 329–342

The Atg8 family: multifunctional ubiquitin-like key regulators of autophagy

Moran Rawet Slobodkin and Zvulun Elazar[1]

Department of Biological Chemistry, The Weizmann Institute of Science, 234 Herzl St, Rehovot 7632700, Israel

Abstract

Autophagy is an evolutionarily-conserved catabolic process initiated by the engulfment of cytosolic components in a crescent-shaped structure, called the phagophore, that expands and fuses to form a closed double-membrane vesicle, the autophagosome. Autophagosomes are subsequently targeted to the lysosome/vacuole with which they fuse to degrade their content. The formation of the autophagosome is carried out by a set of autophagy-related proteins (Atg), highly conserved from yeast to mammals. The Atg8s are Ubl (ubiquitin-like) proteins that play an essential role in autophagosome biogenesis. This family of proteins comprises a single member in yeast and several mammalian homologues grouped into three subfamilies: LC3 (light-chain 3), GABARAP (γ-aminobutyric acid receptor-associated protein) and GATE-16 (Golgi-associated ATPase enhancer of 16 kDa). The Atg8s are synthesized as cytosolic precursors, but can undergo a series of post-translational modifications leading to their tight association with autophagosomal structures following autophagy induction. Owing to this feature, the Atg8 proteins have been widely served as key molecules to monitor autophagosomes and autophagic activity. Studies in both yeast and mammalian systems have demonstrated that Atg8s play a dual role in the autophagosome formation process, coupling between selective incorporation of autophagy cargo and promoting autophagosome membrane expansion and closure. The membrane-remodelling activity of the Atg8 proteins is associated with their capacity to promote tethering and hemifusion of liposomes *in vitro*.

[1]To whom correspondence should be addressed (email zvulun.elazar@weizmann.ac.il).

Keywords:
Atg8, autophagosome, GABARAP, GATE-16, LC3, lysosome, phosphatidylethanolamine, ubiquitin, vacuole.

Introduction

Autophagy is a catabolic process initiated by the sequestration of cytoplasmic proteins and entire organelles within a double membrane sac called the isolation membrane or phagophore. The phagophore expands and is ultimately closed resulting in the formation of the mature autophagosome. The outer membrane of the autophagosome is subsequently fused with the cellular lytic compartment (lysosome in mammals, vacuole in yeast) where the inner membrane together with the sequestered content, termed the autophagic body, is released and degraded by hydrolytic enzymes (Figure 1).

Autophagy was originally implicated in adaptive response of cells to provide nutrients and energy on exposure to stress conditions; however, it has since been connected to diverse physiological and pathological processes including differentiation, development, neurodegeneration, immune function, cancer and aging [1].

Two independent molecular genetic studies in yeast have led to the identification of the autophagy-defective *apg* and *aut* mutants [2,3]. A third overlapping group of mutants (the *cvt* mutants) were subsequently isolated on the basis of defects in the mechanistically related cvt (cytoplasm-to-vacuole-targeting) pathway [4]. The cvt pathway mediates the transport of vacuolar resident enzymes such as the precursor form of API (aminopeptidase-I; proAPI), Ams1 (α-mannosidase) and Ape4 (aspartyl aminopeptidase) from the cytosol to the vacuole under

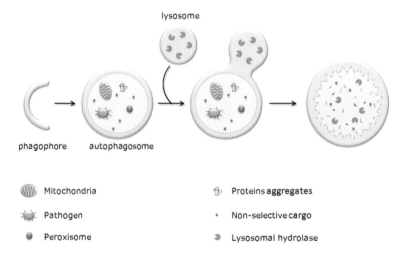

Figure 1. The autophagy process
Autophagy begins with the formation of a double-membrane cup-shaped structure called the phagophore. Cytosolic components including soluble and aggregated proteins, whole organelles and invading pathogens are sequestered within the expanding phagophores to yield mature autophagosomes. Finally, the outer membrane of the autophagosome fuses with the lysosome/vacuole and the released content is degraded by lysosomal hydrolases.

nutrient-rich conditions. The nomenclature of the autophagy-related genes was subsequently unified as ATG [5]. Within the past few years, analysis of the ATG gene products has progressed rapidly and many of their mammalian homologues have been identified and characterized.

The Atg8 family of proteins, constituting a single member in yeast and multiple homologues in higher eukaryotes, were shown to be essential regulators of autophagy [6]. These were the first molecules found to be specifically associated with autophagy-related membrane structures. Atg8 proteins are Ubl (ubiquitin-like) proteins, but rather than conjugating to another protein, they are attached to PE (phosphatidylethanolamine) leading to their tight membrane association. Newly synthesized Atg8 is rapidly C-terminally cleaved by a cysteine protease of the Atg4 family. This cleavage exposes a glycine residue which is then covalently conjugated to PE in a reaction catalysed by the E1- and E2-like enzymes, Atg7 and Atg3 respectively.

Atg8 proteins appear to play multiple roles in the autophagy process. They are responsible for specific recruitment of cargo proteins destined for lysosomal degradation and promote autophagosome maturation. This latter role is suggested to be mediated by their interaction with different effectors involved in regulating the basic autophagy machinery. In addition, Atg8s may directly support membrane fusion events driving the expansion and closure of autophagosomes.

The Atg8 family

Although yeast express only one Atg8 protein, the mammalian Atg8 homologues constitute a family of proteins subdivided into three subfamilies on the basis of amino-acid-sequence homology: MAP1LC3 (microtubule-associated protein 1 light-chain 3; hereafter referred to as LC3), GABARAP (γ-aminobutyric acid receptor-associated protein) and GATE-16 (Golgi-associated ATPase enhancer of 16 kDa). The subfamilies in different organisms vary in their gene numbers due to gene duplication and lost events that occurred during evolution. Humans have a single GATE-16 gene, two GABARAP genes [GABARAP and GABARAPL1 (γ-aminobutyric acid receptor-associated protein-like 1)] and four LC3 genes (LC3A, LC3B, LC3B2 and LC3C) [6].

LC3 was the first mammalian Atg8 homologue to be characterized. It was originally identified as the light chain of the microtubule-associated proteins 1A and 1B in rat brain. Its first implication in autophagy was provided by Kabeya et al. [7] who showed that LC3 is post-translationally processed into two forms: LC3-I, located at cytosolic fractions, and PE-conjugated LC3-II, which associates with the autophagosome membrane. The amount of LC3-II was found to be correlated with the extent of autophagosome formation. The other Atg8 homologues, belonging to the GABARAP and GATE-16 subfamily, were initially characterized as intracellular trafficking factors. GATE-16 was found to interact with the Golgi v-SNARE (vesicle-associated N-ethylmaleimide-sensitive factor attachment protein receptors) GOS-28 as well as with NSF (N-ethylmaleimide-sensitive factor) leading to activation of the ATPase activity of the latter. Through these interactions, GATE-16 was shown to modulate intra-Golgi transport by coupling between NSF activity and SNARE activation [8]. In addition, GATE-16 was implicated in post-mitotic Golgi re-assembly [9]. GABARAP was identified as a cytosolic factor regulating the intracellular transport of the γ2 subunit of $GABA_A$ receptors [10]. It was later shown that, similar to LC3, both GATE-16 and GABARAP are subjected to post-translational processing and exist in two modified forms, I and II [11]. Conversion into the II form correlates with their association with the autophagosome.

© 2013 Biochemical Society

Structural studies of several Atg8 family members indicated that Atg8 proteins share a strong structural similarity to ubiquitin [12]. In addition to the ubiquitin core, comprising five β-strands flanked by two α-helices, Atg8 proteins contain two N-terminal α-helices which play a crucial role in the functionality of these proteins, probably by regulating protein–protein interactions.

Despite their structural resemblance, mammalian Atg8s show some differences in amino acid sequence. For example, the first α-helix in LC3 is basic, whereas in GATE-16 and GABARAP it has an acidic nature. The surface of the second α-helix is acidic in LC3, neutral in GATE-16 and basic in GABARAP. Whereas conserved domains among the mammalians Atg8s are likely to be responsible for characteristic interactions of these proteins, such as the binding to the conjugation machinery proteins, structural differences between the various Atg8 proteins may confer differential specificity towards target proteins and reflect their diverse functions. Indeed, the mammalian homologues of Atg8 were shown to have diverse interaction profiles and to regulate different steps along the autophagic process. Moreover, some differences in the tissue distribution of Atg8s point to a tissue-specific function for some family members [6]. For simplicity, 'Atg8' will hereafter refer to all family members. The specific names will be used to distinguish between yeast Atg8 and the mammalian homologues.

Atg8 processing

The function of the Atg8 proteins is associated with their membrane-binding state. Atg8s are found in cells either as a free cytosolic form, designated Atg8-I, or a tightly membrane-bound form termed Atg8-II. Atg8-II is localized both to the inner and the outer membrane of the autophagosome. The recruitment of Atg8 on to the autophagosomal membrane depends on a series of post-translational modifications (Figure 2). Nascent Atg8 is proteolytically cleaved at its C-terminal end by a constitutively expressed cysteine protease Atg4. Mammalians contain four Atg4 homologues (named Atg4A–D or autophagin1–autophagin4), showing sequence similarity to a single Atg4 in yeast [13]. This proteolytic event produces the mature Atg8-I which possesses a C-terminally exposed glycine residue. Atg8-I is then covalently conjugated to PE through an amide bond between the C-terminal glycine residue and the amino group of PE. PE-conjugated Atg8 (Atg8-II) is associated with the autophagosomal membrane. The lipidation reaction is mediated by a ubiquitination-like system involving the sequential action of Atg7 and Atg3, which serve as E1-activating- and E2-conjugating-like enzymes respectively [14]. Finally, Atg4 can attack Atg8 associated with the outer membrane of the autophagosome, deconjugating it from its target lipid. This makes the association of Atg8 with the autophagosomal membrane a reversible process. The liberated Atg8 is probably recycled and participates in a new conjugation reaction, whereas Atg8–PE entrapped inside the autophagosome is degraded following fusion with the lysosome. Both the conjugating and the deconjugating activities of Atg4 are required for normal progression of autophagy. The recruitment of Atg8 to the autophagosome is preceded by a second Ubl system mediating the conjugation between Ubl, Atg12 and Atg5 [14]. Homologous with the ubiquitin system, Atg12 is first activated by the E1-like enzyme, Atg7, followed by its transfer to an E2-like enzyme, Atg10. Finally, the C-terminal glycine residue of Atg12 is covalently conjugated to an internal lysine residue of Atg5 through an isopeptide bond. The Atg12–Atg5 conjugate then associates with Atg16 which can form homo-oligomers through a coiled-coiled domain. This allows Atg16 to cross-link

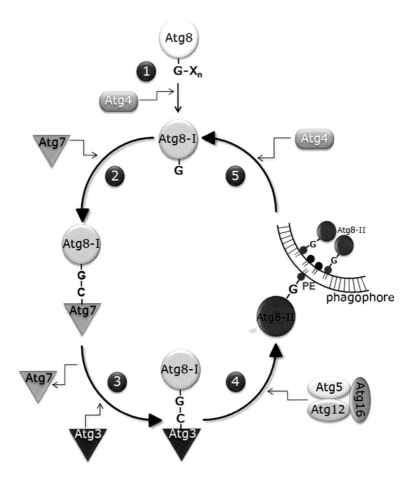

Figure 2. Processing of Atg8s
Following their synthesis as C-terminally extended proforms, Atg8 homologues are cleaved by Atg4 cysteine proteases, resulting in the formation of the Atg8-I form harbouring an exposed C-terminal glycine residue (G) (**1**). The exposed glycine residue forms a thioester bond with a catalytic cysteine residue (C) of the E1-like enzyme Atg7 (**2**). Activated Atg8 is then transferred to the E2-like enzyme Atg3, also through a thioester bond (**3**). Finally, the C-terminal glycine residue of Atg8 is conjugated to PE through an amide bond, resulting in the formation of the tightly membrane-associated Atg8-II form (**4**). Membrane association of Atg8 is reversible as Atg8–PE, which is associated with the outer autophagosome membrane, can be cleaved by Atg4 to release free Atg8-I (**5**).

multiple Atg12–Atg5 conjugates into a single large protein complex. It has been shown that the localization of the Atg12–Atg5–Atg16 complex dictates the site of Atg8 lipidation and facilitates the lipidation reaction by promoting the transfer of Atg8 from Atg3 to PE thus acting as a ubiquitin ligase (E3)-like enzyme [15].

Recruitment of Atg8 on to the autophagosome may be controlled by additional post-translational modifications. LC3 was found to be subjected to PKA (protein kinase A)-mediated phosphorylation [16]. It was further demonstrated that LC3 was dephosphorylated following autophagy induction in neuronal cells. Membrane association of dephosphorylated LC3 was enhanced, whereas formation of LC3-postivie autophagosomes was inhibited by the expression of a pseudophosphorylated LC3 mutant. Finally, acetylation of Atg8 and other Atg proteins by the acetyltransferase p300 was proposed to negatively regulate their activity [17].

Function of the Atg8 family
Autophagosome biogenesis
Initial observations in yeast showing that autophagosome formation was severely impaired in Atg8-null mutants, implicated Atg8 as a crucial player in the process of autophagosome biogenesis [18]. Subsequent findings using both yeast and mammalian systems have highlighted the importance of the Atg8 proteins in the regulation of autophagy. Expression of Atg8 is induced under starvation in yeast. The amount of lipidated Atg8, localized to autophagosomal structures, is greatly increased following autophagy induction and is correlated with the extent of autophagosome formation. The levels of Atg8 determine the levels of autophagy and control the size of autophagosomes [19].

Analysis of the Atg8 role in mammals was initially hampered owing to the existence of multiple homologues. Overexpression of a catalytically dead Atg4B (Atg4B^{C74A}) in mammalians cells was found to inhibit autophagy by sequestration of non-lipidated Atg8 homologues [20]. These cells, overexpressing mutant Atg4B, accumulated unclosed pre-autophagosomal structures, suggesting that lipidation of Atg8 homologues is required for completion of autophagosomes in mammalian cells. Manipulation in the expression of the Atg8 subfamilies in mammals (LC3, GABARAP and GATE-16) by siRNA-mediated knockdown and overexpression approaches revealed that each subfamily contributes to autophagosome biogenesis, but act in different steps along this process. Although the LC3 subfamily mediates the elongation of the phagophore membrane, GABARAP and GATE-16 families may be involved in a downstream step along the maturation process possibly in sealing of the autophagosomes [21].

The first mechanistic insight into the function of Atg8 was provided by Ohsumi and colleagues [22] using an *in vitro* system to reconstitute Atg8–PE conjugation in the presence of liposomes. They showed that lipidated Atg8 mediated the tethering and hemifusion of liposomes to which it was anchored. The study showed further that Atg8 mutants defective in membrane tethering and hemifusion activities impaired the formation of autophagosomes in yeast. These observations indicated a role of Atg8 in membrane remodelling, driving the growth and maturation of autophagosomal structures. It was subsequently shown that the two Atg8 mammalian homologues, LC3 and GATE-16, promote tethering and membrane fusion. This activity is mediated by positively charged and hydrophobic amino acids in the N-terminal α-helices of LC3 and GATE-16 respectively [23].

In addition to the crucial role of Atg8–PE in autophagy, it has been proposed that Atg4-mediated delipidation of Atg8, namely the release of Atg8 from the autophagosome membrane, is an important step required to facilitate normal autophagy. Prevention of Atg8 delipidation was shown to result in its mislocalization to the vacuolar membrane and to have negative effects on autophagosome biogenesis reflected by a reduction in both the number and size of autophagosomes [24]. A possible explanation to this finding is that Atg8 conjugation to PE occurs not only at pre-autophagosomal structures, but also in other sites within the cell. The release of Atg8–PE from these sites is crucial to supply the growing demand for Atg8 following autophagy induction. Deconjugation of Atg8–PE appears to be additionally required in later stages, facilitating autophagosome maturation and fusion with the vacuole [25].

Sorting of autophagy cargo

Autophagy has long been considered to be a non-selective bulk degradation process. Recent experimental data, however, present mounting evidence for another mode of autophagy responsible for selective delivery of cargo for lysosomal degradation [26]. A wide range of substrates found to be specifically cleared by autophagy include: protein inclusions caused by aggregate-prone or misfolded proteins (aggrephagy), organelles such as peroxisomes (pexophagy), mitochondria (mitophagy) and surplus ER (endoplasmic reticulum) (reticulophagy), bacteria and virus (xenophagy), and ribosomes (ribophagy) [26].

In addition to their role in autophagosome biogenesis, Atg8 proteins appear to be central factors in mediating selective cargo sorting into autophagosomes. This activity is largely achieved by interaction with adaptor proteins, also called autophagy receptors, that link autophagy substrates to autophagosome-associated Atg8s. Mostly, the autophagy receptors are themselves degraded by autophagy [26]. In fact, direct Atg8-mediated cargo sorting was first demonstrated in the autophagy-related cvt pathway in yeast. Atg8 was found to interact with Atg19, a receptor for the cvt cargoes facilitating their vacuolar targeting. The cargo-sorting function of mammalian LC3 was subsequently revealed by studies demonstrating its role in the clearance of ubiquitinated molecules. In addition to its fundamental role in protein degradation by the proteasome system, ubiquitin has emerged as a selective degradation signal for the lysosomal targeting of various types of autophagy substrates such as protein aggregates, membrane-bound organelles and microbes [27]. Clearance of such ubiquitinated substrates depends on a group of UBD (ubiquitin-binding domain)-containing autophagy receptors which simultaneously bind to the Atg8 proteins [26].

The rapidly expanding list of autophagy receptors in mammals include: p62/SQSTM1 and neighbour of Brca1 (Nbr1), which are involved in the autophagic degradation of ubiquitinated protein aggregates; p62/SQSTM1, NDP52 (nuclear dot protein 52 kDa) and optenurin, which link LC3 to intracellular pathogens; Nix, the outer mitochondrial membrane protein Bcl2-related protein, which may be directly involved in the recruitment of the autophagy machinery to damaged mitochondria; and autophagy-linked FYVE protein (ALFY). The interaction of autophagy receptors with Atg8 homologues is mediated by a short linear motif termed LIR (LC3-interacting region), conforming to the consensus W/F/Y-X-X-L/I/V and surrounded by at least one acidic residue.

Recruitment of autophagy machinery

LIR-dependent interactions of Atg8 with autophagy core proteins that may serve a regulatory function have also been reported. A direct LIR-mediated binding of Atg8s to the essential autophagy regulator Atg1 (in yeast) and the ULK1 (uncoordinated-51-like kinase 1) complex (in mammals) was found to facilitate autophagosome formation [28,29]. Likewise, Atg8 was found to interact with several members of the TBC (Tre2, Bub2 and Cdc16) domain-containing GAPs (GTPase-activating proteins) of Rab proteins [30]. Rab-type small GTPases are evolutionarily conserved membrane-trafficking proteins, some of which were implicated in regulating tethering and fusion of autophagy-related membranes. Thus, in addition to their

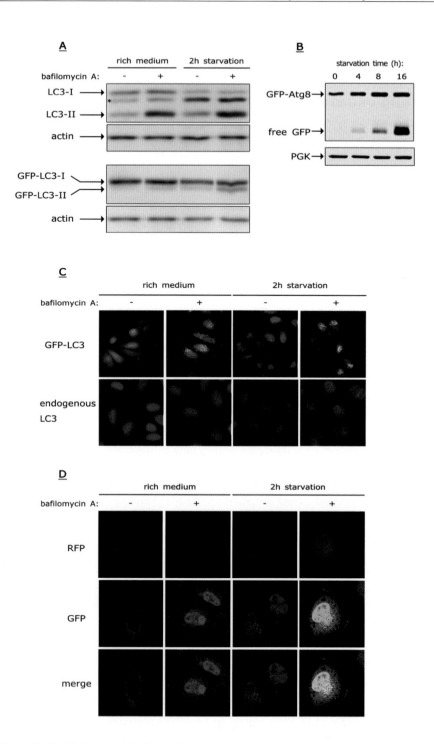

Figure 3. *(See facing page for legend)*

role in recognition of degradation targets, Atg8s may function as scaffold proteins promoting the assembly of critical autophagy complexes on the surface of autophagosomes.

LC3 as a tool to monitor the autophagic process

As Atg8s are specifically associated with premature as well as completed autophagosomes, and since conjugated Atg8 correlates with the number of autophagosomes, these proteins are widely used as specific markers to monitor autophagosomes and autophagic activity. Multiple assays, utilized to detect and quantify Atg8 proteins in yeast and mammals, include Western blotting, fluorescence microscopy and flow cytometry [31,32].

Induction of autophagy is accompanied by an increased conversion of soluble Atg8-I into PE-conjugated membrane-bound Atg8-II. Thus accumulation of Atg8-II should be indicative of autophagy induction. Lipidated Atg8 can be separated from the non-lipidated form owing to its faster migration by electrophoresis to an apparently lower M_r position (albeit it is a larger molecular mass). However, since Atg8-II, associated with the inner autophagosome membrane, is degraded following fusion of the limiting membrane with the lysosome/vacuole, a decrease in Atg8-II levels is often seen under autophagy-inducing conditions. Therefore to reliably measure autophagic activity, Atg8 levels should be determined in the presence or absence of lysosomal degradation inhibitors (Figure 3A). In yeast, the discrimination between the non-lipidated and the lipidated forms of Atg8 can be more complicated due to their nearly identical SDS/PAGE motilities. The more commonly employed assay to follow autophagy in this organism is the GFP–Atg8 processing assay. This assay relies on the use of an ectopically expressed

Figure 3. Analysing autophagy using Atg8 proteins
(**A**) Upper blot shows HeLa cells that were incubated for 2 h in nutrient-rich or in starvation medium in the presence (+) or absence (−) of the lysosomal inhibitor bafilomycin A1. Cells were then lysed and subjected to Western blot analysis using an antibody directed against LC3. Lower blot shows HeLa cells, stably expressing GFP-LC3B, that were similarly treated. Western blot analysis was done with a GFP-directed antibody. The asterisk indicates a non-related band. Note that both in the absence and in the presence of the lysosomal inhibitor, starvation results in a higher amount of lipidated LC3 (LC3-II) compared with the non-lipidated form (LC3-I), suggesting an increase in autophagosome formation. On lysosomal inhibition, the LC3-II/LC3-I ratio is significantly augmented due to the block in lysosome-mediated consumption of autophagosome-associated LC3. Actin served as a loading control. Notably, the LC3 mobility shift can be quantified and used as a measurement for autophagic activity [32]. (**B**) Yeast cells expressing GFP–Atg8 were starved of nitrogen for the indicated times and analysed by Western blotting using an anti-GFP antibody (for details see main text). PGK (phosphoglycerate kinase) served as a loading control. (**C**) Upper panel shows HeLa cells that were treated as in (**A**) and analysed using direct fluorescent microscopy following staining with DAPI to visualize nuclei. Lower panel shows cells stably expressing GFP-LC3B that were similarly treated, fixed and stained with an LC3-directed primary antibody followed by a rhodamine-conjugated secondary antibody. Nuclei were stained with DAPI. (**D**) HeLa cells transiently expressing tandem fluorescent LC3 (RFP–GFP–LC3) were treated as in (**A**) and visualized by fluorescence microscopy. Note the increase in LC3-positive puncta following autophagy induction by starvation. In the absence of lysosomal inhibitor, a higher amount of red puncta (representing autophagosomes already fused with the lysosome) are observed compared with green/yellow puncta (representing the autophagosomal structure before fusion). On addition of the lysosome inhibitor there is a marked increase in the number of LC3-associated autophagosomes which appear as yellow dots due to the block in the degradative activity of the lysosome.

N-terminal GFP-tagged Atg8 (not C-terminally tagged, as the C-terminus of Atg8 is proteolytically processed by Atg4). Following autophagy induction, GFP–Atg8 decorating the inner autophagosome membrane is released into the vacuole lumen. Although Atg8 is rapidly degraded by vacuolar proteases, the relatively stable GFP remains intact leading to the accumulation of free GFP. Thus accumulation of free GFP indicates an increase in the autophagy flux (Figure 3B).

The GFP-tagged Atg8/LC3 can also be used as functional markers for monitoring autophagy through direct fluorescence microscopy in cultured cells and transgenic organisms. Alternatively, endogenous Atg8 proteins can be detected in immunocytochemistry or immunohistochemistry procedures using specific antibodies. By following the endogenous protein, one obviates the need for transfection or generation of a transgenic organism and avoids potential artefacts resulting from overexpression. Atg8-associated autophagic membranes and autophagosome are visualized as bright cytosolically scattered puncta (Figure 3C). It should be noted that accumulation of intracellular Atg8-positive dots may not be indicative of autophagy induction, but rather reflects autophagy inhibition due to blockage in autophagosome consumption (caused by impaired trafficking to lysosomes, compromised autophagosome–lysosome fusion or reduction in lysosomal degradative activity). These possibilities can be distinguished by using lysosomal degradation inhibitors. An increase in the number of Atg8 puncta in the presence of such inhibitors would reflect active autophagy whereas unchanged puncta levels on lysosomal inhibition would be suggestive of autophagy blockage. Another fluorescence approach, capable of analysing both autophagy induction and flux without the necessity for drug treatment, makes use of the tandem monomeric RFP–GFP-tagged Atg8. Whereas the GFP signal is sensitive to the acidic and/or proteolytic conditions of the lysosome lumen, RFP is more stable. Therefore a yellow signal (derived from colocalization of the GFP and then RFP fluorophores) corresponds to autophagosomal structures before their fusion with the lysosome. Autophagosomes, already fused with the lysosome, would appear as red puncta resulting from the RFP, without GFP, signal (Figure 3D). Lastly, FACS (fluorescence-activated cell sorting) has been used for the analysis of autophagy in living mammalian cells. Autophagy induction leads to a lysosomal-activity-dependent decrease in the total cellular signal of fluorescently tagged Atg8. This simple approach allows precise automated analysis of a large number of cells to obtain robust data.

It is important to point out that although the Atg8 family is routinely used to study autophagy, the analysis methods on the basis of these proteins may suffer from various caveats. The amount and the processing of the Atg8s may be tissue- and cell context-dependent. In addition, Atg8s may be involved in non-autophagy cellular processes. Thus it is recommended to include additional tools, such as EM (electron microscopy), and more direct approaches to measure autophagic flux by monitoring autophagic substrate degradation.

Conclusions and future aspects

Since the term 'autophagy' was first coined in 1963 by Christian de Duve, the study of this important cellular process has significantly expanded. Autophagy is now recognized as a crucial pathway involved in normal physiological processes as well as in pathological conditions. In addition to its role in maintaining cellular homoeostasis, autophagy is implicated in a growing list of diseases including cancer, metabolic and neurodegenerative disorders, infectious and inflammatory diseases, diabetes and obesity. Therefore understanding the cellular and molecular mechanisms underlying autophagy is of great interest. Such an improved knowledge would enable identification of new targets for both diagnostic and therapeutic implications.

The Atg8 proteins are key regulators of the autophagy process. Owing to their localization on autophagosomes and intermediate structures, these proteins also serve as reliable markers to monitor autophagy in various biological systems. Although the Atg8 proteins were extensively investigated in the last decade, mechanisms of their action and regulation are not yet fully resolved.

Atg8s were shown to promote proper autophagosome biogenesis. This activity was mechanistically associated with their lipidation-dependent capacity to mediate tethering and fusion of liposomes *in vitro*. A plausible model to explain how these functions of Atg8s mediate autophagosome maturation is that the growth of the autophagosome depends on the supply of lipids and/or proteins provided by vesicular structures. Atg8s associated with these structures as well as with autophagosomal intermediates may drive their fusion leading to the expansion of autophagosomes. Alternatively, Atg8-mediated membrane fusion may promote the final step of autophagosome sealing. Of note, the fundamental cellular membrane fusion molecules, soluble SNAREs, were proposed recently to mediate membrane fusion events along the autophagy process [33]. Whether Atg8s act independently or assist in these SNARE-dependent fusion events needs further study.

Another open question in the field relates to the regulation of Atg8s processing namely, the lipidation/de-lipidation cycle. On the one hand, association of Atg8s with autophagosomal structures is essential for proper autophagosome biogenesis suggesting that there must be a mechanism that protects lipidated Atg8 and prevents unregulated deconjugation. On the other hand, several reports indicate that Atg4-mediated delipidation of Atg8s is also required for normal autophagy. It is not known at what step of autophagosome biogenesis Atg8 deconjugation occurs (following autophagosome completion or earlier). Atg8 processing is likely to be subjected to complex temporal and spatial regulations and, indeed, regulation of Atg4 was documented in mammalian cells. Atg4 activity was shown to be inhibited by the accumulation of ROS (reactive oxygen species), preventing the release of Atg8 proteins from the autophagosomal membrane during autophagy [34].

The Atg8 family emerge as multifunctional autophagy regulators promoting recruitment of degradation targets as well as supporting autophagosome biogenesis by acting as 'fusogens' and by interacting with additional autophagy effectors. However, whether and how these functions of Atg8s are coordinated is not entirely clear. Ho et al. [35] have reported that impaired binding of cargo receptor by Atg8 negatively affects its general autophagy regulation function, suggesting that Atg8 roles in autophagosome formation and cargo sorting are coupled.

Summary

- Autophagy is a catabolic pathway for the delivery of a wide range of substrates including protein aggregates, cellular organelles and invading pathogens to lysosomal degradation through their entrapment in double-membrane structures called autophagosomes.
- Atg8 family members are essential components of the core autophagy machinery.

- The Atg8 family comprises a single member in yeast, whereas in higher organisms it constitutes several homologues subdivided into three subfamilies: LC3, GABARAP and GATE-16.
- Atg8s exist as three forms in cells. They are produced as a C-terminally extended precursor (Atg8) which is rapidly cleaved by the Atg4 cysteine protease to yield the cytosolic Atg8-I form with an exposed C-terminal glycine residue. Following autophagy induction, Atg8-I is converted into the Atg8-II form which tightly associates with autophagosomal membranes through its conjugation to the lipid PE.
- Conversion of Atg8-I into lipidated Atg8-II depends on the sequential action of the E1-activating-like enzyme (Atg7), E2-conjugating-like enzyme (Atg3) and the Atg12–Atg5–Atg16 complex acting as the E3-ubiquitin ligase-like enzyme.
- Atg8s can be dissociated from the autophagosome membrane through their Atg4-mediated deconjugation from PE.
- Atg8s were shown to act as membrane modifiers promoting the expansion and closure of maturing autophagosomes. In addition, Atg8s play a central role in selective recruitment of autophagy substrates into autophagosomes thus mediating their lysosomal targeting and degradation.

Z.E. is the incumbent of the Harold Korda Chair of Biology and is grateful for funding from the Legacy Heritage Fund [grant number 1309/13], the Israeli Science Foundation ISF [grant number 535/2011], the German Israeli Foundation GIF [grant number 1129-157] and the Louis Brause Philanthropic Fund.

References

1. Choi, A.M.K., Ryter, S.W. and Levine, B. (2013) Autophagy in human health and disease. N. Engl. J. Med. **368**, 651–662
2. Tsukada, M. and Ohsumi, Y. (1993) Isolation and characterization of autophagy-defective mutants of *Saccharomyces cerevisiae*. FEBS Lett. **333**, 169–174
3. Thumm, M., Egner, R., Koch, B., Schlumpberger, M., Straub, M., Veenhuis, M. and Wolf, D.H. (1994) Isolation of autophagocytosis mutants of *Saccharomyces cerevisiae*. FEBS Lett. **349**, 275–280
4. Harding, T.M., Morano, K.A., Scott, S.V. and Klionsky, D.J. (1995) Isolation and characterization of yeast mutants in the cytoplasm to vacuole protein targeting pathway. J. Cell Biol. **131**, 591–602
5. Klionsky, D.J., Cregg, J.M., Dunn, Jr, W.A., Emr, S.D., Sakai, Y., Sandoval, I.V., Sibirny, A., Subramani, S., Thumm, M., Veenhuis, M. and Ohsumi, Y. (2003) A unified nomenclature for yeast autophagy-related genes. Dev. Cell **5**, 539–545
6. Shpilka, T., Weidberg, H., Pietrokovski, S. and Elazar, Z. (2011) Atg8: an autophagy-related ubiquitin-like protein family. Genome Biol. **12**, 226
7. Kabeya, Y., Mizushima, N., Ueno, T., Yamamoto, A., Kirisako, T., Noda, T., Kominami, E., Ohsumi, Y. and Yoshimori, T. (2000) LC3, a mammalian homologue of yeast Apg8p, is localized in autophagosome membranes after processing. EMBO J. **19**, 5720–5728
8. Sagiv, Y., Legesse-Miller, A., Porat, A. and Elazar, Z. (2000) GATE-16, a membrane transport modulator, interacts with NSF and the Golgi v-SNARE GOS-28. EMBO J. **19**, 1494–1504

9. Muller, J.M., Shorter, J., Newman, R., Deinhardt, K., Sagiv, Y., Elazar, Z., Warren, G. and Shima, D.T. (2002) Sequential SNARE disassembly and GATE-16-GOS-28 complex assembly mediated by distinct NSF activities drives Golgi membrane fusion. J. Cell Biol. **157**, 1161–1173
10. Leil, T.A., Chen, Z.W., Chang, C.S. and Olsen, R.W. (2004) GABAA receptor-associated protein traffics GABAA receptors to the plasma membrane in neurons. J. Neurosci. **24**, 11429–11438
11. Kabeya, Y., Mizushima, N., Yamamoto, A., Oshitani-Okamoto, S., Ohsumi, Y. and Yoshimori, T. (2004) LC3, GABARAP and GATE16 localize to autophagosomal membrane depending on form-II formation. J. Cell Sci. **117**, 2805–2812
12. Sugawara, K., Suzuki, N.N., Fujioka, Y., Mizushima, N., Ohsumi, Y. and Inagaki, F. (2004) The crystal structure of microtubule-associated protein light chain 3, a mammalian homologue of *Saccharomyces cerevisiae* Atg8. Genes Cells **9**, 611–618
13. Mariño G, Uría JA, Puente XS, Quesada V, Bordallo J, López-Otín C (2003) Human autophagins, a family of cysteine proteinases potentially implicated in cell degradation by autophagy. J. Biol. Chem. **278**, 3671–3678
14. Geng, J. and Klionsky, D.J. (2008) The Atg8 and Atg12 ubiquitin-like conjugation systems in macroautopahgy. 'Protein modifications: beyond the usual suspects' review series. EMBO Rep. **9**, 859–864
15. Hanada, T., Noda, N.N., Satomi, Y., Ichimura, Y., Fujioka, Y., Takao, T., Inagaki, F. and Ohsumi, Y. (2007) The Atg12–Atg5 conjugate has a novel E3-like activity for protein lipidation in autophagy. J. Biol. Chem. **282**, 37298–37302
16. Cherra, 3rd, S.J., Kulich, S.M., Uechi, G., Balasubramani, M., Mountzouris, J., Day, B.W. and Chu, C.T. (2010) Regulation of the autophagy protein LC3 by phosphorylation. J. Cell Biol. **190**, 533–539
17. Lee, I.H. and Finkel, T. (2009) Regulation of autophagy by the p300 acetyltransferase. J. Biol. Chem. **284**, 6322–6328
18. Kirisako, T., Baba, M., Ishihara, N., Miyazawa, K., Ohsumi, M., Yoshimori, T., Noda, T. and Ohsumi, Y. (1999) Formation process of autophagosome is traced with Apg8/Aut7p in yeast. J. Cell Biol. **147**, 435–446
19. Xie, Z., Nair, U. and Klionsky, D.J. (2008) Atg8 controls phagophore expansion during autophagosome formation. Mol. Biol. Cell **19**, 3290–3298
20. Fujita, N., Hayashi-Nishino, M., Fukumoto, H., Omori, H., Yamamoto, A., Noda, T. and Yoshimori, T. (2008) An Atg4B mutant hampers the lipidation of LC3 paralogues and causes defects in autophagosome closure. Mol. Biol. Cell **19**, 4651–4659
21. Weidberg, H., Shvets, E., Shpilka, T., Shimron, F., Shinder, V. and Elazar, Z. (2010) LC3 and GATE-16/GABARAP subfamilies are both essential yet act differently in autophagosome biogenesis. EMBO J. **29**, 1792–1802
22. Nakatogawa, H., Ichimura, Y. and Ohsumi, Y. (2007) Atg8, a ubiquitin-like protein required for autophagosome formation, mediates membrane tethering and hemifusion. Cell **130**, 165–178
23. Weidberg, H., Shpilka, T., Shvets, E., Abada, A., Shimron, F. and Elazar, Z. (2011) LC3 and GATE-16 N termini mediate membrane fusion processes required for autophagosome biogenesis. Dev. Cell **20**, 444–454
24. Nair, U., Yen, W.L., Mari, M., Cao, Y., Xie, Z., Baba, M., Reggiori, F. and Klionsky, D.J. (2012) A role for Atg8–PE deconjugation in autophagosome biogenesis. Autophagy **8**, 780–793
25. Yu, Z.Q., Ni, T., Hong, B., Wang, H.Y., Jiang, F.J., Zou, S., Chen, Y., Zheng, X.L., Klionsky, D.J., Liang ,Y. and Xie, Z. (2012) Dual roles of Atg8–PE deconjugation by Atg4 in autophagy. Autophagy **8**, 883–892
26. Johansen, T. and Lamark, T. (2011) Selective autophagy mediated by autophagic adaptor proteins. Autophagy **7**, 279–296
27. Shaid, S., Brandts, C.H., Serve, H. and Dikic, I. (2013) Ubiquitination and selective autophagy. Cell Death Differ. **20**, 21–30
28. Nakatogawa, H., Ohbayashi, S., Sakoh-Nakatogawa, M., Kakuta, S., Suzuki, S.W., Kirisako, H., Kondo-Kakuta, C., Noda, N.N., Yamamoto, H. and Ohsumi, Y. (2012) The autophagy-

related protein kinase Atg1 interacts with the ubiquitin-like protein Atg8 via the Atg8 family interacting motif to facilitate autophagosome formation. J. Biol. Chem. **287**, 28503–28507
29. Alemu, E.A., Lamark, T., Torgersen, K.M., Birgisdottir, A.B., Larsen, K.B., Jain, A., Olsvik, H., Øvervatn, A., Kirkin, V. and Johansen, T. (2012) ATG8 family proteins act as scaffolds for assembly of the ULK complex: sequence requirements for LC3-interacting region (LIR) motifs. J. Biol. Chem. **287**, 39275–39290
30. Popovic, D., Akutsu, M., Novak, I., Harper, J.W., Behrends, C. and Dikic, I. (2012) Rab GTPase-activating proteins in autophagy: regulation of endocytic and autophagy pathways by direct binding to human ATG8 modifiers. Mol. Cell. Biol. **32**, 1733–1744
31. Cheong, H. and Klionsky, D.J. (2008) Biochemical methods to monitor autophagy-related processes in yeast. Methods Enzymol. **451**, 1–26
32. Klionsky, D.J., Abeliovich, H., Agostinis, P., Agrawal, D.K., Aliev, G., Askew, D.S., Baba, M., Baehrecke, E.H., Bahr, B.A., Ballabio, A. et al. (2008) Guidelines for the use and interpretation of assays for monitoring autophagy in higher erkaryotes. Autophagy **4**, 151–175
33. Stroupe, C. (2011) Autophagy: cells SNARE selves. Curr. Biol. **21**, R697–R699
34. Scherz-Shouval, R., Shvets, E., Fass, E., Shorer, H., Gil, L. and Elazar, Z. (2007) Reactive oxygen species are essential for autophagy and specifically regulate the activity of Atg4. EMBO J. **26**, 1749–1760
35. Ho, K.H., Chang, H.E. and Huang, W.P. (2009) Mutation at the cargo-receptor binding site of Atg8 also affects its general autophagy regulation function. Autophagy **5**, 461–471

6

Autophagosome maturation and lysosomal fusion

Ian G. Ganley[1]

MRC Protein Phosphorylation Unit, College of Life Sciences, University of Dundee, Dow Street, Dundee DD1 5EH, Scotland, U.K.

Abstract

Compartmentalization is essential in the eukaryotic cell and this is most often achieved by sequestering specific components that perform a related function in a membrane-bound organelle. To function normally these organelles must transiently fuse with other compartments in order to transfer protein and lipid that is needed for them to function. These events must be highly coordinated otherwise non-specific fusion could occur leading to loss of compartment identity and function. The autophagosome is a specialized membrane compartment that delivers cytosolic components to the lysosome for degradation. Likewise, this delivery is coordinated so that only when the autophagosome is fully formed is it imparted with the information to allow it to specifically fuse with the endocytic system and deliver its contents to the lysosome. In the present chapter, I discuss our current understanding of how this happens.

Keywords:
actin, autophagosome, endosome, fusion, lysosome, maturation, microtubule, Rab, SNARE, tether.

Introduction

Lysosomes are the degradative organelles of the cell and contain an acidic lumen to allow specialized hydrolases to break down protein, lipid, carbohydrate and nucleic acids [1,2]. Material to be degraded can come from the outside, cell surface or inside of the cell and is delivered to

[1]email i.ganley@dundee.ac.uk

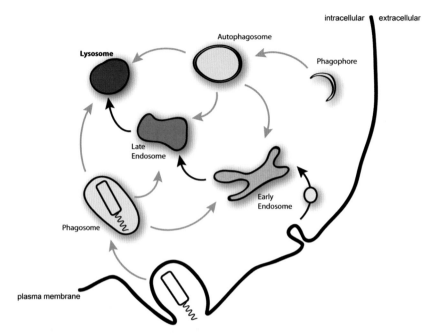

Figure 1. Summary of lysosomal degradation pathways
Lysosomes are the digestive organelles of the cell and the source of the material to be degraded determines which pathway is taken. Intracellular material is taken up by the autophagic pathway (shown at the top), whereas small extracellular molecules and cell-surface receptors are delivered to the lysosome by the endocytic pathway (shown in the centre). In a similar fashion, bulk uptake of extracellular material (pinocytosis) or pathogens proceeds via the phagocytic pathway (shown at the bottom). Once material has been sequestered, delivery to the lysosome proceeds by a similar mechanism for all the pathways, with autophagosomes and phagosomes essentially merging with endosomes en route to lysosomes. Hence the endocytic pathway is central to all the lysosomal degradation pathways.

the lysosome in a series of inter-related transport pathways (Figure 1). As lysosomes are the final destination, inhibition of lysosomal function will disrupt all of these pathways. Of these, the endocytic pathway is the most important as it is central to lysosomal biogenesis and function; the pathway supplies not only the hydrolases, but also proteins such as the vacuolar ATPases and glycosylated LAMPs that acidify and allow the organelle to cope with its digestive lumenal environment respectively. Given this, the endocytic pathway is also central to autophagy. The endocytic pathway itself starts at the cell surface, where external solutes and plasma membrane components are internalized and delivered to a series of sorting compartments where the majority of protein and membrane is recycled. Components such as hydrolases and those destined for destruction are further sorted to the lysosome (for a review see [3]). Disruption of this pathway at multiple points leads to impaired lysosomal function and hence autophagosome turnover. Surprisingly this is not only due to the accumulation of undegraded autophagic components in the lysosome, but also because the autophagosomes themselves fail to fuse with the lysosome. For example, disruption of early endosomal function by depletion of COPI (coatomer protein I) (see Table 1 for a brief description of proteins discussed below) leads to the accumulation of autophagic structures that fail to reach the lysosome [4]. Likewise, loss of function of the ESCRT (endosomal sorting complex required for transport) that is required for

Table 1. Glossary of proteins implicated in autophagosome–lysosome fusion

Protein	Function in endocytosis and autophagy
COPI	Vesicle coat protein complex involved in endosomal and Golgi sorting and early endosomal fusion with autophagosomes
ESCRT	Series of protein complexes involved in sorting cargo into multivesicular bodies and required for autophagosome fusion
ATG8	Family of ubiquitin-like modifiers, including LC3, that are essential for autophagosome maturation
Rab5	Small GTPase involved in early endosomal function and autophagosome formation
Rab7	Small GTPase involved in endosomal/autophagosome maturation
Rab33b	Small GTPase involved in autophagosome formation
OATL1/TBC1D25	Rab33b GAP
RILP	Rab7 effector protein that binds to dynein motors
FYCO1	Rab7 effector that binds to kinesins
HDAC6	Recruits actin to aid in specific autophagosome fusion
HOPS	Endosome–autophagosome tethering complex [six subunits: VPS11, VPS16 (interacts with UVRAG), VPS18, VPS33, VPS39 (Rab7 GEF) and VPS41]
VPS34	Class III PI3P required for autophagy and endocytosis
UVRAG	Activator of VPS34 and VPS16 interactor
Rubicon	Inhibitor of UVRAG and VPS34
TECPR1	Autophagosomal-lysosomal tether
Syntaxin 7, syntaxin 8, Vti1b	Q-SNAREs implicated in endosome–lysosome fusion
VAMP7, VAMP8	R-SNAREs implicated in endosome–lysosome fusion
Syntaxin 17, SNAP-29	Q-SNAREs implicated in autophagosome–lysosome fusion
VAMP3, VAMP8	R-SNAREs implicated in autophagosome–lysosome fusion

generation of intraluminal vesicles in later endosomal structures not only blocks endocytic degradation, but also leads to the accumulation of autophagosomes [5]. Taken together, this highlights the importance of endocytosis in autophagy. These data also raise another important question: where exactly does the autophagosome fuse with the endocytic system? It seems likely that autophagosomes can fuse with the endocytic system at multiple points. By fusing with both early and late endosomes they form a hybrid organelle termed an amphisome, which then goes on to fuse with lysosomes to form autolysosomes. It is, however, possible that autolysosomes are formed directly by autophagosome–lysosome fusion. It should also be noted that the endocytic pathway is not just important for autophagosomal maturation and turnover, but as discussed in Chapter 3, is also important for autophagosome biogenesis.

Not only is the autophagosomal fusion step essential for autophagy, it also must be taken into account when measuring the process experimentally; an increase in autophagosomes does not necessarily mean increased autophagy as it might equally result from a block in autophagosome fusion. This means that as well as autophagosome formation, lysosomal fusion and turnover must also be measured: the so-called autophagic flux. A very elegant method to visualize flux in cells was developed by Yoshimori and co-workers [6] in an assay that relies on a tandem-tagged RFP–GFP–LC3 (light-chain 3) (Figure 2). The chemical properties of these two fluorophores mean that in the cytosol and on autophagosomes both will fluoresce; however, on autophagosome fusion with lysosomes the acidic environment quenches the GFP signal, but not the RFP signal. Therefore immature autophagosomes will fluoresce both red and green, whereas autolysosomes will only fluoresce red.

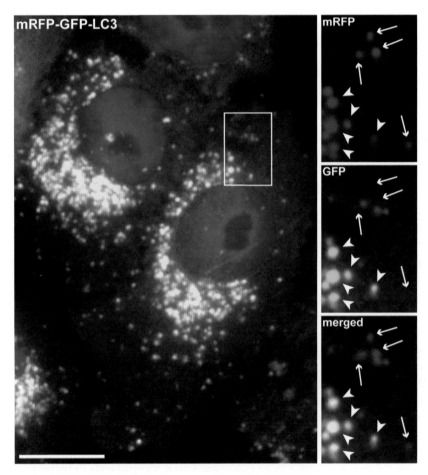

Figure 2. Visualizing autophagosomes fusing with lysosomes
Fluorescence micrograph showing immortalized human breast epithelial cells expressing a tandem mRFP (red) and GFP (green)-tagged version of the autophagosomal marker LC3. Cells have been deprived of amino acids for 2 h to trigger autophagy, as visualized by LC3 puncta. Early autophagosomes and phagophores are positive for both green and red fluorescence (examples are indicated by arrowheads in the magnified right-hand panels). However, once autophagosomes have fused with lysosomes to form autolysosomes, the GFP signal is quenched and the LC3 fluoresces red only (examples are indicated by arrows in the right-hand panels). The nuclei of the cells are stained blue using DAPI and the scale bar is 10 μm.

Maturation of the autophagosome

Once the autophagosome has formed it must travel to the endocytic system to fuse and deliver its contents for degradation. The timing of this is very important and must only happen once the phagophore has sealed – if the fusion machinery is recruited and activated before this, then the cytosolic cargo will be released or left attached to the cytosolic side of the lysosomal membrane, unable to be degraded. Therefore these fusion factors must only recognize the mature autophagosome and how this is achieved is currently unknown. One potential signal may come from the ATG8 family of proteins. As discussed in Chapter 5, this family of ubiquitin-like proteins is thought to be involved in sealing the phagophore to form the autophagosome. Work from yeast has suggested that loss of Atg8 from the outer autophagosomal membrane is a signal for fusion competency by allowing removal of the autophagy initiating machinery [7,8]. Indeed, immunoelectron microscopy suggests that the majority of Atg8 is present on the inner membrane of mature autophagosomes [9]. It is not clear whether a similar requirement of ATG8 removal exists in mammalian cells, especially as some forms, such as LC3, may be required for movement of autophagosomes towards lysosomes (see below). However, proteins associated with the phagophore, such as ATG16 and the ULK1 (uncoordinated-51-like kinase 1) complex, are absent from the mature autophagosome, implying that they are removed before fusion with the lysosome. This mechanism and whether it is a prerequisite for fusion is currently unknown.

Travelling to endosomes and lysosomes: role of the cytoskeleton

The cytoskeleton has many cellular functions that range from structural maintenance of cell shape to cell division and movement. Of relevance, microtubules and actin filaments affect membrane transport in the following manner: these structures form an interconnected network that act as highways on which vesicles and organelles, with the aid of motor proteins, move along to reach their destinations. It is therefore no surprise that microtubules and actin filaments have been implicated in the movement of autophagosomes and fusion with the endocytic system (Figure 3) [10].

Microtubules

Microtubules are a dynamic polymer consisting of α- and β-tubulin dimers that form a hollow cylindrical structure. Microtubules are polar with two distinct ends: a minus end that is often attached to the centrosome, and a plus end that stretches towards the periphery of the cell. The plus end is the more dynamic of the two and can undergo rapid cycles of polymerization/depolymerization. Movement of protein or organelle cargo along the surface of microtubules is driven by the kinesin and dynein motor protein families, which are powered by ATP hydrolysis, permitting them to move in a stepwise fashion [11]. Kinesins are plus-end-directed motors; they transport their cargoes towards the plus end of microtubules, whereas dyneins are minus-end-directed motors. In this way, movement can be directed either towards the periphery or centre of the cell.

Surprisingly, microtubules appear to be dispensable for autophagy in yeast. Forced microtubule depolymerization using the chemical nocodazole or deletion of the *tub2* gene (β-tubulin)

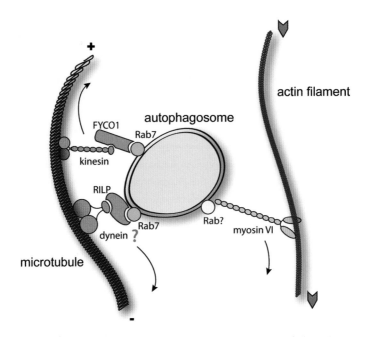

Figure 3. Movement of autophagosomes along the cytoskeleton
Rab7 GTPase links autophagosomes to microtubule motors through FYCO1 to mediate kinesin-driven movement towards the cell periphery, and possibly through RILP to allow interaction with dynein and movement towards the cell centre. The pointed-end directed actin motor, myosin VI, has also been linked with autophagosome movement, although whether this is mediated by a Rab GTPase is unknown.

does not affect bulk autophagy and fusion with the vacuole [9]. However, the situation appears slightly different in mammalian cells and although microtubules may not be absolutely essential, they likely facilitate this fusion step [10]. Autophagosomes are thought to form randomly throughout the cytoplasm, yet late endosomes and lysosomes are predominantly perinuclear. Therefore movement of autophagosomes to the proximity of these organelles will increase fusion efficiency and, indeed, live-cell microscopy has shown that mature autophagosomes move along microtubule tracks [12–15]. Perhaps the most striking evidence for this has come from studies in primary neurons that shows autophagosomes form distally, but migrate inwards along the axon in a microtubule- and dynein-dependent manner until fusion with endosomes and lysosomes [16]. The positive role of dynein has also been linked to autophagosomal fusion in studies showing that its loss leads to accumulation of LC3-II and reduction in the autophagic clearance of aggregate-prone proteins [17].

Given the more perinuclear distribution of late endosomes and lysosomes, kinesins that move in the opposite direction may counteract the actions of dynein. However, it is likely that both operate in combination to optimize movement of autophagosomes, given that depletion of the kinesin KIF5B blocked autophagy and resulted in tight perinuclear clustering of autophagosomes [18]. It is interesting to note that localization of lysosomes themselves can determine the rate of autophagosomal fusion. Increasing the perinuclear localization of lysosomes by depletion of the kinesins KIF1B-β and KIF2A led to increased autophagosomal fusion, whereas dispersion of lysosomes to the periphery by overexpression of the motors reduced autophagosomal fusion [19].

How then are autophagosomes connected to microtubule proteins? An important player is the small GTPase Rab7. Rabs act as molecular switches and are master coordinators of multiple stages of membrane trafficking (see below and [20]). Rab7 localizes to late endosomes and lysosomes and is involved in their motility and fusion. For example, Rab7 recruits dynein motors to endosomes through the interaction of its effector protein RILP (Rab-interacting lysosomal protein) [21]. Given that Rab7 appears to be essential for autophagosomal fusion with lysosomes [22], it is possible, although evidence is currently lacking, that a similar mechanism with respect to RILP recruitment occurs with autophagosomes. Rab7 has also been linked to recruitment of kinesin motors to autophagosomes from a study on the protein FYCO1 (FYVE and coiled-coil domain containing 1) [23]. FYCO1 is recruited to membranes by direct interaction with GTP-bound Rab7 as well as interacting with the lipid PI3P (phosphatidylinositol 3-phosphate). FYCO1 also contains an LC3-interacting motif that, in combination with Rab7 and PI3P binding, specifically recruits it to autophagosomes. The study also showed that overexpression or depletion of FYCO1 redistributed autophagosomes to the periphery or cell centre respectively, consistent with a role in kinesin recruitment (Figure 3).

Actin filaments

Actin filaments comprise polymerized actin monomers and, similar to microtubules, this polymerization is dynamic. However, actin filaments are shorter and appear to have a more random orientation, although they do have polarity with filament ends termed either barbed or pointed. As with microtubules, actin filaments can act as tracks to move various intracellular cargoes via the myosin family of motor proteins [24].

Evidence suggests that actin is involved in selective autophagy induction in yeast, but there are few data to suggest involvement in fusion of autophagosomes with the vacuole [10]. In contrast, actin may be involved in mammalian autophagosomal fusion. Recent work has demonstrated that HDAC6 (histone deacetylase 6) helps to recruit cortactin, a protein that can activate actin polymerization, to autophagosomes involved in clearing protein aggregates [25]. The authors found that loss of HDAC6, cortactin or actin polymerization led to a block in the fusion of autophagosomes with lysosomes. Surprisingly, this block was specific to autophagosomes targeting protein aggregates as general starvation-induced autophagy was unaffected. This is an important observation as it suggests that not all autophagosomes are made equal; the itinerary of the autophagosome could be dependent on the cargo it contains. Actin motors have also been implicated as loss of myosin VI leads to the accumulation of autophagosomes [26]. The authors also showed that myosin VI interacts with Tom1, a component of the ESCRT machinery, suggesting that the actin motor is important for fusion with ESCRT-containing endosomes (Figure 3).

Fusion of the autophagosome

Once the autophagosome has arrived at its destination then it must fuse with the endocytic system. Our knowledge of the machinery involved in this process stems from our understanding of general intracellular membrane trafficking events, which involve the coordination of three sets of protein families: Rab GTPases, membrane-tethering complexes and SNAREs (soluble *N*-ethylmaleimide-sensitive factor-attachment protein receptor). In this way, specific fusion events can be coordinated as Rab proteins localize to specific membranes and recruit

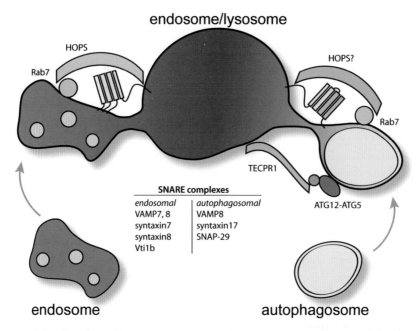

Figure 4. Endosome and autophagosome fusion
Endosome fusion with endosomes/lysosomes mirrors that of autophagosomal fusion. First, a long-range docking interaction occurs between a tethering complex. For endosomes this is the HOPS complex that interacts with Rab7 on the incoming endosome and possibly the R-SNARE VAMP7 on the target endosome/lysosome. This long-range interaction allows the *trans*-SNARE complex to form, which drives the actual fusion event. Given Rab7's involvement in autophagosomal fusion it is also possible that the HOPS complex is the autophagosome–lysosome tether. An additional or alternate tether maybe the protein TECPR1, which localizes to lysosomes and also interacts with the ATG12–ATG5 complex on the outer autophagosomal membrane. The SNARE complex required for autophagosomal fusion does however differ from the endosomal complex, with the requirement for the Q-SNAREs syntaxin 17 and SNAP29, although the R–SNARE on the target membrane may be the same (VAMP8).

tethering molecules that act as bridges to bring the compartments intended for fusion into close proximity. The tethers in turn prime SNARE proteins that physically drive the fusion of opposing lipid bilayers. This is discussed in more detail below and summarized in Figure 4.

Rabs: master coordinators of membrane trafficking

Rabs are the largest family of small GTPases with over 60 members in humans and regulate multiple stages of membrane trafficking. They recruit specific effector proteins such as cargo adaptors into forming transport vesicles, motor proteins to allow movement of vesicles to their target membrane and tethers to aid the fusion machinery in delivery of vesicular cargo when it reaches its destination [20]. Rabs localize to distinct membrane compartments and in this way they are thought to provide specificity in membrane trafficking. Membrane-associated Rabs are activated by specific GEFs (guanine nucleotide-exchange factors) that drive GTP binding. On binding GTP Rabs undergo a conformational change that allows interaction with its effector proteins. Rabs are in turn inactivated by a specific GAP (GTPase-activating protein) that stimulates hydrolysis of the bound GTP to GDP causing loss of effector binding and extraction from membranes.

As well as being involved in autophagosome movement, Rab7 also has a more direct role in the fusion event. Rab7 is essential for late endosomal/lysosomal fusion in conjunction with the multi-subunit HOPS (homotypic fusion and vacuolar protein sorting) tethering complex (see below and [3]). It must be remembered, however, that blockage of endocytosis will also indirectly block autophagosomal fusion. Therefore a factor may not be involved in autophagy, yet still cause a block. Given Rab7's involvement in endosome maturation and fusion, it could be one such factor. However, evidence suggests a more direct role for Rab7 in autophagy. A recent study showed that the compound thapsigargin (an inhibitor of the SERCA calcium pump) blocked recruitment of Rab7 to mature autophagosomes and inhibited their fusion with the endocytic system, yet did not block recruitment to endosomes, or block endosomal fusion [27]. This implies a specific role for Rab7 in autophagy. A critical factor therefore is likely to be recruitment of the Rab protein to the mature autophagosome. How this happens with respect to autophagosomes is currently unknown, but is likely to involve the coordination of multiple Rab effector proteins and exchange factors [20]. An intriguing possibility is through a process called Rab conversion, where an upstream Rab recruits a GEF that enhances association and activation of a downstream Rab. In this way, maturation of a membrane compartment can be coordinated and has been elegantly shown in the switch of early endosomes to late endosomes [28,29]. Here, early endosomal Rab5 recruits the HOPS complex, the VPS39 subunit of which acts as a GEF to activate Rab7. In turn, negative Rab5 feedback occurs and the endosome matures to a Rab7-positive Rab5-negative structure that can then go on to fuse with the lysosome. As Rab7 has been linked to autophagosomal fusion, a similar mechanism could operate here, and in support of this, Rab5 has been linked to autophagy induction [30,31]. However, there could be an alternate upstream Rab. Rab33b is a Golgi-localized Rab that plays a role in autophagosome formation through interaction with its autophagy effector protein ATG16 [32]. Interestingly, OATL1/TBC1D25 has been identified as a GAP for Rab33b and its overexpression inhibits autophagosomal–lysosomal fusion [33]. It is possible that OATL1 acts in the negative feedback during Rab conversion to allow the autophagosome to mature to a Rab33b-free form. Whether Rab33b interacts with the HOPS complex/Rab7, or an as yet unidentified Rab/tether combination, is currently unknown. Regardless, Rab7 is an important factor in both endosomal and autophagic maturation.

Lipid signalling also plays an important role in Rab activation and membrane fusion. The PI3K (phosphoinositide 3-kinase) VPS34 catalyses the formation of PI3P and is important for endosomal Rab5–Rab7 conversion mentioned previously [29]. VPS34 exists in multiple complexes within the cell, but one complex containing the protein UVRAG (UV radiation resistance-associated gene) is directly linked to endosomal and autophagic maturation. UVRAG stimulates VPS34 and also binds to VPS16, a subunit of the HOPS complex. The UVRAG interaction with VPS16 is thought to aid in membrane recruitment of the HOPS complex which further activates Rab7 through its VPS39 GEF subunit [34]. In support of this, the protein Rubicon, which binds to UVRAG and negatively regulates VPS34 activity, inhibits endosomal and autophagosome fusion with lysosomes. Interestingly, active Rab7 displaces the UVRAG–Rubicon interaction, freeing up UVRAG to bind to HOPS complex that, in turn, further activates Rab7 [35,36]. This provides a strong feed-forward signal to drive the fusion machinery.

Membrane tethers: bridges to fusion

Membrane tethers are thought to provide another level of specificity and facilitate the docking and fusion process by bridging the opposing membranes and/or stimulating SNARE complex formation. As mentioned, Rab proteins function to recruit and activate tethering factors, of which there are two broad types: long coiled-coil domain-containing proteins, or multi-subunit tethering complexes [37]. As discussed previously, the HOPS complex is thought to act as a tether between late endosomes and lysosomes. Although direct evidence is lacking for a HOPS role in autophagosomal fusion, it is an attractive candidate given its interaction with Rab7. The HOPS complex falls into the multi-subunit family of tethers and contains six subunits linking Rab7 activation through its VPS39 subunit, as well as interaction with UVRAG through its VPS16 subunit [37]. In yeast, the HOPS complex has also been shown to bind to and 'proofread' the vacuolar SNAREs Vam3p, Vam7p and Vti1p to enhance efficiency of fusion [38]. Whether this is the case with the mammalian complex is not clear, neither is it clear if this only happens in endosomal–lysosomal fusion or can also be applicable to autophagosomal–lysosomal fusion. Regardless, by interacting with Rab7 on the incoming autophagosome/endosome and the SNARE protein on the target endosome/lysosome, the HOPS complex bridges the two compartments to facilitate their fusion.

In addition to the HOPS complex, another molecule has been highlighted recently that could perform a tethering function in autophagosome fusion with the endocytic system. TECPR1 (tectonin β-propeller repeat-containing protein 1) was identified as a protein that interacts with the autophagy-initiating protein ATG5 during bacterial infection, implying a role in specific autophagosome formation [39]. However, recent studies have shown that TECPR1 localizes to mature autophagosomes and lysosomes and its loss leads to autophagosome accumulation. Interestingly TECPR1 binds to PI3P, the product of the VPS34 reaction, in an ATG5–ATG12-dependent manner. This PI3P binding appears essential for TECPR1 function [40]. Whether TECPR1 interacts with any SNARE or Rab proteins remains to be seen.

SNAREs: the driving force of membrane fusion

SNAREs are membrane-anchored proteins localized on opposing membrane compartments that can interact with each other to form a highly energetically favourable complex. It is the energy released on forming this complex that is thought to provide the driving force behind membrane fusion. SNARES are essential for vesicular transport and cell function, this is highlighted by the fact that the most toxic protein known to man, botulinum toxin, is an inhibitor of the neuronal SNARE SNAP-25 (25 kDa synaptosome-associated protein). For SNAREs to drive membrane fusion they must form a *trans*-SNARE complex consisting of one R-SNARE [with a key arginine (R) residue in the SNARE motif] on the donor membrane and three Q-SNAREs [with a key glutamine (Q) residue in the SNARE motif] on the acceptor membrane [41]. It is the 'zipping-up' into a four-helical bundle of multiple SNARE complexes that brings the opposing lipid bilayers together to allow fusion to occur. The Q-SNAREs syntaxin 7, syntaxin 8 and Vti1b, along with the R-SNAREs VAMP7 and VAMP8, have been linked to the fusion of late endosomes and lysosomes and thus play a role in autophagosome maturation indirectly [42]. It is possible that these SNAREs also function in autophagosomal fusion, indeed VAMP8 and Vti1b, but not VAMP7, syntaxin 7 or syntaxin 8, were shown to be involved in autophagosomal fusion during clearance of intracellular bacteria [43]. In contrast, another study suggested that VAMP3 and VAMP7 are required sequentially to allow autophagosomes to fuse with endosomes and then lysosomes [44]. As these SNAREs are common to the endocytic pathway, interpretation can be difficult due to the essential role of endocytosis in autophagy. However, a recent study has identified a Q-SNARE that is specific for autophagy. Syntaxin 17 is recruited to autophagosomes and its loss disrupts autophagosome fusion with lysosomes, but not endosome fusion [45]. This work also demonstrated that syntaxin 17 interacted with another Q-SNARE, SNAP-29, as well as the R-SNARE VAMP8. In support of these data, work in *Drosophila melanogaster* has also shown that syntaxin 17 is required for autophagosomal fusion with lysosomes [46]. Given that the block in fusion is specific for autophagy, this likely represents a *bone fide* autophagosomal SNARE complex. However, more work is needed to clarify this in light of a recent study suggesting that syntaxin 17 can operate at an earlier step in autophagosome biogenesis [47].

Conclusion

Lysosomal fusion is not the final stage of autophagy as the autophagosomal contents must then be degraded and the resultant metabolites transported into the cytosol. However, lysosomes are the terminal point and therefore delivery and fusion of autophagosomes to these organelles is critical. As discussed, much of the machinery required for endosome maturation and fusion is shared with the autophagic system and essentially the mature autophagosome can be considered as a specialized endosome with respect to lysosomal fusion. Even so, there is still a lot of ambiguity with regards to autophagosome maturation. Is it a 'one-size-fits-all' scenario for autophagosomes in terms of fusion with the endocytic pathway? Does the type of autophagy (specific or non-specific), or for example the source of the phagophore membrane, determine at which point in the endocytic system the autophagosome fuses and which Rabs/tethers/SNAREs are utilized? Additionally, autophagosome formation is tightly controlled, especially under stresses such as nutrient starvation; however, it remains to be seen whether the fusion step is also regulated, or is it that once autophagosomes have formed they are essentially on a conveyor belt to the lysosome. A deeper understanding of these events will hopefully allow specific manipulation of the autophagic pathway to treat certain medical conditions, something which is becoming increasingly attractive given the links between autophagy and disease (see later chapters).

Summary
- Lysosomes and hence endocytosis are essential for autophagy.
- Disruption of endocytosis will block autophagy.
- Autophagosomes can fuse with the endocytic system at early and late points.
- Microtubules and actin filaments enhance fusion by bringing autophagosomes, endosomes and lysosomes into close proximity.
- The general rules that govern endocytic fusion likely apply to autophagic fusion: Rabs recruit tethers recruit SNAREs.

- Many components of the endocytic and autophagic fusion machinery are shared such as Rab7 and the HOPS complex, yet certain SNAREs may be unique to one process.
- Specific compared with non-specific autophagosomes may have different fusion itineraries.

I am thankful to the Ganley laboratory for critically reading the chapter and fruitful discussions.

References

1. Luzio, J.P., Pryor, P.R. and Bright, N.A. (2007) Lysosomes: fusion and function. Nat. Rev. Mol. Cell Biol. **8**, 622–632
2. Saftig, P. and Klumperman, J. (2009) Lysosome biogenesis and lysosomal membrane proteins: trafficking meets function. Nat. Rev. Mol. Cell Biol. **10**, 623–635
3. Huotari, J. and Helenius, A. (2011) Endosome maturation. EMBO J. **30**, 3481–3500
4. Razi, M., Chan, E.Y. and Tooze, S.A. (2009) Early endosomes and endosomal coatomer are required for autophagy. J. Cell Biol. **185**, 305–321
5. Rusten, T.E. and Stenmark, H. (2009) How do ESCRT proteins control autophagy? J. Cell Sci. **122**, 2179–2183
6. Kimura, S., Noda, T. and Yoshimori, T. (2007) Dissection of the autophagosome maturation process by a novel reporter protein, tandem fluorescent-tagged LC3. Autophagy **3**, 452–460
7. Yu, Z.Q., Ni, T., Hong, B., Wang, H.Y., Jiang, F.J., Zou, S., Chen, Y., Zheng, X.L., Klionsky, D.J., Liang, Y. and Xie, Z. (2012) Dual roles of Atg8-PE deconjugation by Atg4 in autophagy. Autophagy **8**, 883–892
8. Nair, U., Yen, W.L., Mari, M., Cao, Y., Xie, Z., Baba, M., Reggiori, F. and Klionsky, D.J. (2012) A role for Atg8-PE deconjugation in autophagosome biogenesis. Autophagy **8**, 780–793
9. Kirisako, T., Baba, M., Ishihara, N., Miyazawa, K., Ohsumi, M., Yoshimori, T., Noda, T. and Ohsumi, Y. (1999) Formation process of autophagosome is traced with Apg8/Aut7p in yeast. J. Cell Biol. **147**, 435–446
10. Monastyrska, I., Rieter, E., Klionsky, D.J. and Reggiori, F. (2009) Multiple roles of the cytoskeleton in autophagy. Biol. Rev. Cambridge Philos. Soc. **84**, 431–448
11. Hunt, S.D. and Stephens, D.J. (2011) The role of motor proteins in endosomal sorting. Biochem. Soc. Trans. **39**, 1179–1184
12. Fass, E., Shvets, E., Degani, I., Hirschberg, K. and Elazar, Z. (2006) Microtubules support production of starvation-induced autophagosomes but not their targeting and fusion with lysosomes. J. Biol. Chem. **281**, 36303–36316
13. Kimura, S., Noda, T. and Yoshimori, T. (2008) Dynein-dependent movement of autophagosomes mediates efficient encounters with lysosomes. Cell Struct. Funct. **33**, 109–122
14. Jahreiss, L., Menzies, F.M. and Rubinsztein, D.C. (2008) The itinerary of autophagosomes: from peripheral formation to kiss-and-run fusion with lysosomes. Traffic **9**, 574–587
15. Kochl, R., Hu, X.W., Chan, E.Y. and Tooze, S.A. (2006) Microtubules facilitate autophagosome formation and fusion of autophagosomes with endosomes. Traffic **7**, 129–145
16. Maday, S., Wallace, K.E. and Holzbaur, E.L. (2012) Autophagosomes initiate distally and mature during transport toward the cell soma in primary neurons. J. Cell Biol. **196**, 407–417
17. Ravikumar, B., Acevedo-Arozena, A., Imarisio, S., Berger, Z., Vacher, C., O'Kane, C.J., Brown, S.D. and Rubinsztein, D.C. (2005) Dynein mutations impair autophagic clearance of aggregate-prone proteins. Nat. Genet. **37**, 771–776

18. Cardoso, C.M., Groth-Pedersen, L., Hoyer-Hansen, M., Kirkegaard, T., Corcelle, E., Andersen, J.S., Jaattela, M. and Nylandsted, J. (2009) Depletion of kinesin 5B affects lysosomal distribution and stability and induces peri-nuclear accumulation of autophagosomes in cancer cells. PLoS ONE **4**, e4424
19. Korolchuk, V.I., Saiki, S., Lichtenberg, M., Siddiqi, F.H., Roberts, E.A., Imarisio, S., Jahreiss, L., Sarkar, S., Futter, M., Menzies, F.M. et al. (2011) Lysosomal positioning coordinates cellular nutrient responses. Nat. Cell Biol. **13**, 453–460
20. Stenmark, H. (2009) Rab GTPases as coordinators of vesicle traffic. Nat. Rev. Mol. Cell Biol. **10**, 513–525
21. Jordens, I., Fernandez-Borja, M., Marsman, M., Dusseljee, S., Janssen, L., Calafat, J., Janssen, H., Wubbolts, R. and Neefjes, J. (2001) The Rab7 effector protein RILP controls lysosomal transport by inducing the recruitment of dynein–dynactin motors. Curr. Biol. **11**, 1680–1685
22. Hyttinen, J.M., Niittykoski, M., Salminen, A. and Kaarniranta, K. (2013) Maturation of autophagosomes and endosomes: a key role for Rab7. Biochim. Biophys. Acta **1833**, 503–510
23. Pankiv, S., Alemu, E.A., Brech, A., Bruun, J.A., Lamark, T., Overvatn, A., Bjorkoy, G. and Johansen, T. (2010) FYCO1 is a Rab7 effector that binds to LC3 and PI3P to mediate microtubule plus end-directed vesicle transport. J. Cell Biol. **188**, 253–269
24. Hartman, M.A., Finan, D., Sivaramakrishnan, S. and Spudich, J.A. (2011) Principles of unconventional myosin function and targeting. Annu. Rev. Cell Dev. Biol. **27**, 133–155
25. Lee, J.Y., Koga, H., Kawaguchi, Y., Tang, W., Wong, E., Gao, Y.S., Pandey, U.B., Kaushik, S., Tresse, E., Lu, J. et al. (2010) HDAC6 controls autophagosome maturation essential for ubiquitin-selective quality-control autophagy. EMBO J. **29**, 969–980
26. Tumbarello, D.A., Waxse, B.J., Arden, S.D., Bright, N.A., Kendrick-Jones, J. and Buss, F. (2012) Autophagy receptors link myosin VI to autophagosomes to mediate Tom1-dependent autophagosome maturation and fusion with the lysosome. Nat. Cell Biol. **14**, 1024–1035
27. Ganley, I.G., Wong, P.M., Gammoh, N. and Jiang, X. (2011) Distinct autophagosomal–lysosomal fusion mechanism revealed by thapsigargin-induced autophagy arrest. Mol. Cell **42**, 731–743
28. Rink, J., Ghigo, E., Kalaidzidis, Y. and Zerial, M. (2005) Rab conversion as a mechanism of progression from early to late endosomes. Cell **122**, 735–749
29. Poteryaev, D., Datta, S., Ackema, K., Zerial, M. and Spang, A. (2010) Identification of the switch in early-to-late endosome transition. Cell **141**, 497–508
30. Ravikumar, B., Imarisio, S., Sarkar, S., O'Kane, C.J. and Rubinsztein, D.C. (2008) Rab5 modulates aggregation and toxicity of mutant huntingtin through macroautophagy in cell and fly models of Huntington disease. J. Cell Sci. **121**, 1649–1660
31. Dou, Z., Pan, J.A., Dbouk, H.A., Ballou, L.M., Deleon, J.L., Fan, Y., Chen, J.S., Liang, Z., Li, G., Backer, J.M. et al. (2013) Class IA PI3K p110β subunit promotes autophagy through Rab5 small GTPase in response to growth factor limitation. Mol. Cell **50**, 29–42
32. Itoh, T., Fujita, N., Kanno, E., Yamamoto, A., Yoshimori, T. and Fukuda, M. (2008) Golgi-resident small GTPase Rab33B interacts with Atg16L and modulates autophagosome formation. Mol. Biol. Cell **19**, 2916–2925
33. Itoh, T., Kanno, E., Uemura, T., Waguri, S. and Fukuda, M. (2011) OATL1, a novel autophagosome-resident Rab33B-GAP, regulates autophagosomal maturation. J. Cell Biol. **192**, 839–853
34. Liang, C., Lee, J.S., Inn, K.S., Gack, M.U., Li, Q., Roberts, E.A., Vergne, I., Deretic, V., Feng, P., Akazawa, C. and Jung, J.U. (2008) Beclin1-binding UVRAG targets the class C Vps complex to coordinate autophagosome maturation and endocytic trafficking. Nat. Cell Biol. **10**, 776–787
35. Sun, Q., Westphal, W., Wong, K.N., Tan, I. and Zhong, Q. (2010) Rubicon controls endosome maturation as a Rab7 effector. Proc. Natl. Acad. Sci. U.S.A. **107**, 19338–19343
36. Tabata, K., Matsunaga, K., Sakane, A., Sasaki, T., Noda, T. and Yoshimori, T. (2010) Rubicon and PLEKHM1 negatively regulate the endocytic/autophagic pathway via a novel Rab7-binding domain. Mol. Biol. Cell **21**, 4162–4172

37. Brocker, C., Engelbrecht-Vandre, S. and Ungermann, C. (2010) Multisubunit tethering complexes and their role in membrane fusion. Curr. Biol. **20**, R943-R952
38. Starai, V.J., Hickey, C.M. and Wickner, W. (2008) HOPS proofreads the *trans*-SNARE complex for yeast vacuole fusion. Mol. Biol. Cell **19**, 2500–2508
39. Ogawa, M., Yoshikawa, Y., Kobayashi, T., Mimuro, H., Fukumatsu, M., Kiga, K., Piao, Z., Ashida, H., Yoshida, M., Kakuta, S. et al. (2011) A Tecpr1-dependent selective autophagy pathway targets bacterial pathogens. Cell Host Microbe **9**, 376–389
40. Chen, D., Fan, W., Lu, Y., Ding, X., Chen, S. and Zhong, Q. (2012) A mammalian autophagosome maturation mechanism mediated by TECPR1 and the Atg12-Atg5 conjugate. Mol. Cell **45**, 629–641
41. Jahn, R. and Scheller, R.H. (2006) SNAREs: engines for membrane fusion. Nat. Rev. Mol. Cell Biol. **7**, 631–643
42. Pryor, P.R., Mullock, B.M., Bright, N.A., Lindsay, M.R., Gray, S.R., Richardson, S.C., Stewart, A., James, D.E., Piper, R.C. and Luzio, J.P. (2004) Combinatorial SNARE complexes with VAMP7 or VAMP8 define different late endocytic fusion events. EMBO Rep. **5**, 590–595
43. Furuta, N., Fujita, N., Noda, T., Yoshimori, T. and Amano, A. (2010) Combinational soluble *N*-ethylmaleimide-sensitive factor attachment protein receptor proteins VAMP8 and Vti1b mediate fusion of antimicrobial and canonical autophagosomes with lysosomes. Mol. Biol. Cell **21**, 1001–1010
44. Fader, C.M., Sanchez, D.G., Mestre, M.B. and Colombo, M.I. (2009) TI-VAMP/VAMP7 and VAMP3/cellubrevin: two v-SNARE proteins involved in specific steps of the autophagy/multivesicular body pathways. Biochim. Biophys. Acta **1793**, 1901–1916
45. Itakura, E., Kishi-Itakura, C. and Mizushima, N. (2012) The hairpin-type tail-anchored SNARE syntaxin 17 targets to autophagosomes for fusion with endosomes/lysosomes. Cell **151**, 1256–1269
46. Takats, S., Nagy, P., Varga, A., Pircs, K., Karpati, M., Varga, K., Kovacs, A.L., Hegedus, K. and Juhasz, G. (2013) Autophagosomal Syntaxin17-dependent lysosomal degradation maintains neuronal function in *Drosophila*. J. Cell Biol. **201**, 531–539
47. Hamasaki, M., Furuta, N., Matsuda, A., Nezu, A., Yamamoto, A., Fujita, N., Oomori, H., Noda, T., Haraguchi, T., Hiraoka, Y. et al. (2013) Autophagosomes form at ER–mitochondria contact sites. Nature **495**, 389–393

7

Selective autophagy

Steingrim Svenning and Terje Johansen[1]

Molecular Cancer Research Group, Institute of Medical Biology, University of Tromsø, 9037 Tromsø, Norway

Abstract

During the last decade it has become evident that autophagy is not simply a non-selective bulk degradation pathway for intracellular components. On the contrary, the discovery and characterization of autophagy receptors which target specific cargo for lysosomal degradation by interaction with ATG8 (autophagy-related protein 8)/LC3 (light-chain 3) has accelerated our understanding of selective autophagy. A number of autophagy receptors have been identified which specifically mediate the selective autophagosomal degradation of a variety of cargoes including protein aggregates, signalling complexes, midbody rings, mitochondria and bacterial pathogens. In the present chapter, we discuss these autophagy receptors, their binding to ATG8/LC3 proteins and how they act in ubiquitin-mediated selective autophagy of intracellular bacteria (xenophagy) and protein aggregates (aggrephagy).

Keywords:
aggrephagy, NBR1, NDP52, optineurin, p62, selective autophagy, TBK1, ubiquitin, xenophagy.

Introduction

Eukaryotic cells utilize two mechanistically distinct but largely complementary systems, the UPS (ubiquitin–proteasome system) and the lysosome (or vacuole in yeast and plants), to effectively degrade cellular components. Macroautophagy (hereafter autophagy) is characterized by formation of a double-membrane autophagosome that envelopes components of the cytoplasm and then fuses to a late endosome or lysosome for degradation of its content.

---

[1]To whom correspondence should be addressed (email terje.johansen@uit.no).

During the 1990s genetic screens in yeast identified the *ATG* (autophagy-related protein) genes responsible for this process [1]. Other chapters in this volume describe the molecular players in the autophagosome assembly pathway. Most likely, autophagy evolved both as a non-selective pathway for restoring intracellular nutrient supply during starvation, and as a quality control mechanism to facilitate selective removal of toxic or surplus structures [2].

Lipidated ATG8 [LC3 (light-chain 3) and GABARAP (γ-aminobutyric acid receptor-associated protein) proteins in mammals] is located both on the inner and outer membrane of autophagic vesicles, and acts as a perfect scaffold for specific recruitment of proteins to the phagophore (forming autophagosome). Selective autophagy is based on the recognition and degradation of a specific cargo, in a process depending on receptor proteins that bind ATG8/LC3 to facilitate enrichment of cargoes sequestrated for degradation (Figure 1). Most autophagy receptors bind ATG8/LC3 through a short LIR (LC3-interacting region) motif. The LIR motif is a degenerate sequence with a core motif corresponding to W/F/Y-XX-L/I/V (where X can be any amino acid). They target either ubiquitinated cargoes, interact directly with the cargo, or constitute ATG8-interacting ligands in the outer membrane of mitochondria. Studies of the yeast Cvt (cytoplasm-to-vacuole targeting) pathway, which mediates selective import of two precursor enzymes to the yeast vacuole, have also contributed to the mechanistic understanding of autophagy as a selective process [3]. However, the first selective autophagy receptor to be identified was the human protein p62/SQSTM1 (sequestosome 1) [4,5]. This protein was initially characterized as a scaffold protein acting in several important signalling pathways [6]. In addition, p62 was known to accumulate in ubiquitin-containing protein bodies in a number of protein aggregation diseases [7], creating a link to human pathophysiology that instigated an increased interest in autophagy as a selective process.

Selective autophagy can be classified according to the cargoes involved (Table 1). In the present chapter we focus on the autophagy receptors, their interaction with ATG8/LC3

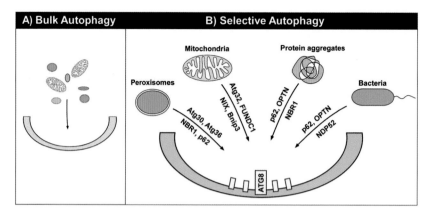

Figure 1. Bulk autophagy versus selective autophagy
(**A**) During non-selective bulk autophagy, cytosolic cargoes are indiscriminately sequestered to the forming autophagosome, the phagophore. (**B**) Selective autophagy enriches the phagophore for specific cargoes in a process dependent on receptor proteins. Several different cargoes have been identified and indicated here along with some of the autophagy receptors commonly associated with selective turnover of these structures including. OPTN, NBR1, NDP52, NIX (NIP3-like protein X) and FUNDC1 (FUN14 domain containing 1).

Table 1. Types of selective autophagy: nomenclature

ER, endoplasmic reticulum.

Name	Cargo	Autophagy receptor(s)
Aggrephagy	Protein aggregates	p62 [5], NBR1 [18] and optineurin [24]
Pexophagy	Peroxisomes	NBR1 [22], p62 [22,45], Atg30 [46] and Atg36 [47]
Mitophagy	Mitochondria	NIX [14], Bnip3 [11], FUNDC1 [13] and Atg32 [12,15]
Xenophagy	Bacteria, virus and protozoans	p62 [39], NDP52 [19] and optineurin [20]
Glycophagy	Glycogen	Stbd1 [16]
ERphagy (reticulophagy)	Parts of the ER	?
Zymophagy	Zymogen granules	?
Lipophagy	Lipid droplets	?
Ribophagy	Ribosomes	?
Nucleophagy	Pieces of the nucleus	?

proteins and their roles in ubiquitin-mediated degradation of intracellular bacteria (xenophagy) and protein aggregates (aggrephagy).

Autophagy receptors

Concomitant with an improved knowledge about cargoes comes identification of novel autophagy receptors orchestrating the process of recognition and delivery. Their function can be outlined by four common criteria: (i) autophagy receptors bind ATG8/LC3 through a conserved binding site, the LIR motif; (ii) autophagy receptors are degraded with the cargo; (iii) deletion of autophagy receptors should not interfere with the function of the basic autophagy machinery; and (iv) protein–protein binding domains or membrane-binding domains mediate interaction with the cargo.

It is important to mention that among a growing number of LIR-containing proteins interacting with ATG8/LC3/GABARAP proteins, many are not autophagy receptors [8].

On the basis of structural features and how they interact with the respective cargoes, autophagy receptors can be divided into the following four different groups (Figure 2): (i) SLRs (sequestosome 1-like receptors) share several similarities structurally and functionally with p62/SQSTM1 [9]. The SLRs all contain dimerization or polymerization domains, bind ubiquitin and interact with members of the ATG8 family proteins. These autophagy receptors recognize ubiquitinated cargoes via C-terminal ubiquitin-binding domains with defined affinities

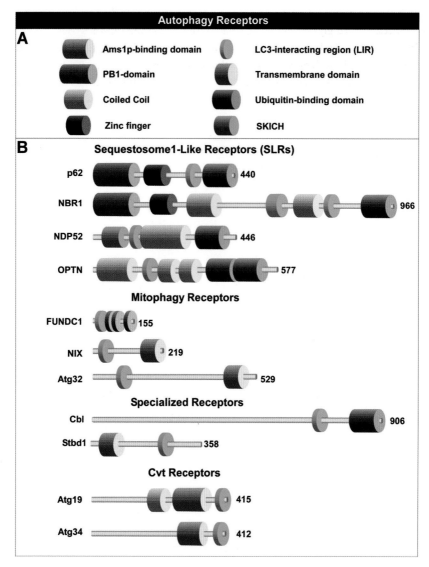

Figure 2. Receptor proteins in selective autophagy
(**A**) Conserved domains important for the function of autophagy receptors. (**B**) Schematic illustration of protein domain architectures of different autophagy receptors. The Cbl protein contains additional domains (from the N-terminus: 4H, EF, SH2 and RING) not shown here for clarity.

for different ubiquitin moieties. However, in some cases they can also bind directly to the unmodified cargo [2,10]. (ii) Autophagic removal of damaged mitochondria (mitophagy) represents the best-studied example of selective autophagy mediated by membrane-associated receptors. Mammalian NIX (NIP3-like protein X), Bnip3, FUNDC1 and yeast Atg32 localize to the mitochondrial outer membrane, where they are anchored by transmembrane domains and mediate selective clearance of mitochondria through LIR-dependent interaction with ATG8 family proteins [11–15] (Figure 2). (iii) Recently, some specialized autophagy receptors have emerged as exemplified by the Stbd1 (starch-binding-domain-containing protein 1) and the

E3-ubiquitin ligase Cbl. Stbd1 is suggested to act as a selective autophagy receptor for glycogen in a process named glycophagy [16] (Table 1). (iv) In the yeast Cvt pathway, the vacuolar hydrolases aminopeptidase 1 (Ape1p) and α-mannosidase (Ams1p) are selectively imported to the yeast vacuole through direct binding to the autophagy receptor Atg19. Binding initiates the formation of a multimeric, or aggregated, Cvt complex. Atg19 then interacts with the adaptor Atg11, which mediates translocation to the forming autophagosome where Atg19 binds to Atg8 to facilitate delivery of the hydrolases [3]. Although Atg19 acts under nutrient-rich conditions, the homologous protein Atg34 acts under nutrient starvation-induced autophagy to transport Ape1p and Ams1p to the vacuole [17].

As mentioned above, p62 was the first selective autophagy receptor described in mammalian cells [4,5]. Kirkin et al. [18], who originally cloned p62, noted the ability of p62 to form aggregates, and coined the name SQSTM1 on the basis of this property. So far, four human ubiquitin-binding autophagy receptors have been described; p62/SQSTM1, NBR1 [neighbour of BRCA1 (breast cancer early-onset 1) gene 1] [18], NDP52 (nuclear dot protein 52) [19] and optineurin [20] (Figure 2).

Human p62 is 440 amino acids long and contains an N-terminal PB1 (Phox and Bem1p) domain, a LIR motif and a C-terminal UBA (ubiquitin-associated) domain (Figure 2). The PB1 domain has oppositely charged surface areas and allows p62 to self-interact, forming polymers in a front-to-back interaction. Polymerization of p62 is essential for selective degradation of p62 by autophagy, and among the human PB1 domain proteins only p62 has this ability. p62 binds ubiquitin via the C-terminal UBA domain. The polymerization- and ubiquitin-binding ability allows p62 to cluster ubiquitinated cargo. The LIR motif mediates interaction with LC3 and is necessary for the degradation of p62 and p62-containing structures by autophagy [5,21]. p62 has been instrumental for the understanding of receptor-mediated autophagy and has been most studied for its role in the clearance of protein aggregates (aggrephagy).

NBR1 and p62 display low sequence similarity, but share related domain architectures. They bind to each other via their PB1 domains, and act as partners in the clearance of protein aggregates, peroxisomes and midbody rings [2,18,22]. NBR1 cannot polymerize via PB1, but self-interacts via a coiled-coil domain. Homologues of NBR1 are found throughout the eukaryotic kingdom, whereas the presence of p62 is unique for metazoans and likely the result of a gene duplication event early in the metazoan lineage [23]. During evolution p62 has lost several domains found in NBR1. Therefore these two proteins may also have some independent roles in selective autophagy. Supporting this, N-terminal to its UBA domain, NBR1 has a small amphipathic α-helix, the J-domain, that mediates interaction with cellular membranes. NBR1 was recently linked to J-domain-dependent degradation of peroxisomes in a process which includes, but does not depend on, p62 [22].

The two other SLRs, optineurin and NDP52, have both been linked to selective autophagy of bacteria (xenophagy) [19,20]. Optineurin has additionally been implicated in ubiquitin-independent selective autophagy of various protein aggregates [24]. The domain architecture of both proteins comprises coiled-coil domains, functional LIR motifs and C-terminal ubiquitin-binding domains (Figure 2). Optineurin has been identified in protein aggregates of several neurodegenerative diseases, and given its ability to self-interact and oligomerize through coiled-coil domains it is likely to function as an aggregating cargo receptor like p62. Optineurin is also involved in ubiquitin-independent aggrephagy of huntingtin aggregates [24]. Less is known about the aggrephagy properties of NDP52, but it has been implicated recently in selective autophagy of the microRNA processing DICER and the effector AGO2 (Argonaute 2) [25].

Interaction of autophagy receptors with ATG8/LC3

The interaction with ATG8 family proteins constitutes the pivotal link between autophagy receptors and the phagophore. Yeast has a single ATG8 protein, *Drosophila melanogaster* and *Caenorhabditis elegans* have two, *Arabidopsis thaliana* has nine and mammals have at least seven. Mammalian ATG8s can be grouped into two subfamilies which includes the LC3s (LC3A, LC3B and LC3C) and the GABARAPs (GABARAP, GABARAPL1 and GABARAPL2). Structurally, ATG8 proteins consist of two N-terminal α-helices and a ubiquitin-like C-terminal domain made up of four-stranded β-sheets with two α-helices on either side (Figure 3A). The ATG8 interaction between LC3B and p62 was mapped to the LIR motif [5]. The functional LIR motif sequence in human p62 stretches from residues 335 to 441 (DDDWTHL). Structural studies of the complex between the p62 LIR peptide and LC3B have shown that the Trp^{338} and Lys^{341} bind in two hydrophobic pockets in the ubiquitin-like domain

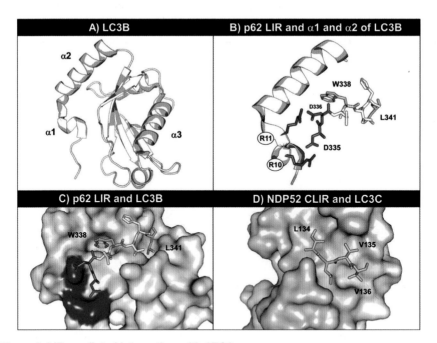

Figure 3. LIR-mediated interaction with ATG8
(**A**) Structural cartoon representation of LC3B. (**B**) α-Helices 1 and 2 constitute the N-terminal arm. Aspartic acid residues in p62–LIR Asp^{337} and Asp^{338} (red) form electrostatic interactions with basic residues Arg^{10} and Arg^{11} (blue) in the N-terminal arm of LC3B. (**C**) Surface representation of LC3B with the bound LIR peptide from p62. Trp^{338} and Leu^{341} in p62-LIR bind in two hydrophobic pockets (bright yellow) in the ubiquitin-like domain of LC3B. (**D**) Surface representation of LC3C with the non-canonical LIR motif of NDP52 bound to a hydrophobic patch on LC3C (bright yellow). The structural data are from the PDB entries 2ZJD and 2VVW.

of LC3B. The first two aspartic acid residues in the LIR motif form electrostatic interactions with basic residues Arg^{10} and Arg^{11} in the N-terminal arm of LC3B (Figures 3B and 3C) [21]. Following the initial characterization of the LIR motif in p62, similar motifs have been identified in numerous other ATG8 interactors, and a consensus LIR motif can be written as W/F/Y-X_1X_2-L/I/V, with an aromatic amino acid in the first position and a bulky hydrophobic amino acid in the fourth position. Acidic residues are frequently found in one to three positions preceding the aromatic residue. If there are no acidic residues at these positions there is usually an acidic residue at X_1, like in yeast Atg19 [26].

A bioinformatics search for a short sequence like the LIR motif will yield a high number of candidate sequences, and a previous proteomics study identified 67 proteins that interact with the human ATG8 orthologues [27]. An LIR motif alone is not sufficient for recruitment of a protein to the inner surface of the phagophore. Analogous to the oligomerization and aggregation of pre-Ape1 in the Cvt pathway, PB1-domain-mediated polymerization of p62 is absolutely required for its degradation by autophagy. In higher eukaryotes, such as mammals and plants, the occurrence of multiple ATG8 homologues adds additional layers of complexity to the regulation of selective autophagy since different autophagy receptors may bind to different ATG8 proteins. Supporting this notion, an atypical LIR motif has been identified in NDP52, composed of three consecutive hydrophobic amino acids (LVV) that recognize a hydrophobic patch on LC3C (Figure 3D). Only LC3C provides the right structure to compensate for the lack of aromatic residues in NDP52–LIR [28]. In future studies it will be important to determine the exact roles of different ATG8 homologues and their interaction with autophagy receptors.

Ubiquitin-mediated degradation

Ubiquitin modifications are involved in mediating cargo recognition by autophagy receptors in both aggrephagy and xenophagy. Ubiquitin is a small 76-amino-acid protein used as a secondary modifier by covalent attachment to other cellular proteins. Post-translational modification by ubiquitin acts as a versatile cellular signal controlling a wide range of biological processes including cell signalling, trafficking and the DNA damage response. Analogous to the proteasome, where ubiquitinated proteins are delivered by ubiquitin receptors, ubiquitin modification is also involved in determining cargo specificity in selective autophagy. Ubiquitination is mediated by a series of enzymatic reactions involving activating (E1), conjugating (E2) and ligating (E3) enzymes generating a covalent bond between the exposed C-terminal glycine residue in ubiquitin and a lysine residue in the cargo protein. Ubiquitin has seven internal lysine residues that can be linked to form polyubiquitin chains (Lys^6, Lys^{11}, Lys^{27}, Lys^{29}, Lys^{33}, Lys^{48} and Lys^{63}), and conjugation on the N-terminal methionine (Met^1) leads to formation of linear chains (Figure 4). The two most prevalent linkages are Lys^{48}- and Lys^{63}- linked chains. The linkage determines the structure of the ubiquitin chain, and signal specificity is achieved by alternative interactions with ubiquitin-binding proteins [10]. Traditionally implied in the context of UPS-mediated protein degradation, Lys^{48}-linked chains adopt compact conformations with the interaction surface partially buried within the fold of the chain (Figure 4). In contrast, Lys^{63}- and Met^1-linked chains adopt open conformations and are associated with autophagy as well as non-proteolytic pathways. Both chain conformations have been implicated in xenophagy [29], whereas Lys^{63}-linked chains have been suggested to be important in aggrephagy [30,31].

© 2013 Biochemical Society

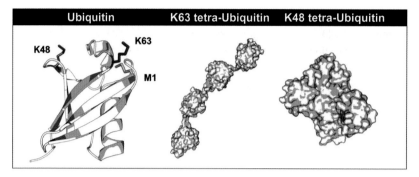

Figure 4. Topology of ubiquitin chains
Structure of the ubiquitin molecule with the internal Lys[48] and Lys[63] highlighted (blue) and the N-terminal Met[1] indicated (yellow). Coupling through Lys[63] creates an extended-structure chain, whereas Lys[48]-linked chains adopt a compact conformation (PDB entries 1UBQ, 3HM3 and 2O6V).

Selective autophagy of protein aggregates: aggrephagy

Mutations, incomplete translation, aberrant protein modifications or failing complex formations can give rise to misfolded proteins. Most cells have a constitutive need for selective autophagy as a protein quality control mechanism acting together with the UPS to degrade misfolded proteins. These have an inherent tendency to aggregate owing to exposed hydrophobic patches that are normally concealed in the native folded state. Aggregation may compromise the role of functional proteins that are sequestered to the aggregate, and lead to a cascading problem for the cell. Actually, large protein aggregates are probably inert to the UPS leaving selective autophagy as the only available degradation route. The main source of post-translational damage to proteins is caused by ROS (reactive oxygen species) formed as natural by-products of normal metabolism of oxygen. Persistent or extensive oxidative damage promotes protein aggregation. The activity of different intracellular proteolytic systems decreases with aging, and deficient removal of oxidized proteins causes accumulation of toxic protein aggregates. This is a hallmark of several common neurodegenerative diseases such as Alzheimer's and Parkinson's diseases, and protein misfolding disorders or proteinopathies in general [7,30]. The aggregates associated with these types of diseases will often contain intrinsically disordered proteins that are prone to aggregation, such as the hyperphosphorylated tau-containing neurofibrillary tangle in Alzheimer's disease brains, aggregated α-synuclein in the Lewy body in Parkinson's disease brains and the intracellular inclusion of the N-terminal fragments of mutant huntingtin in Huntington's disease brains [32].

Mice deficient for autophagy show tissue-specific accumulation of p62-positive structures containing ubiquitinated protein aggregates [33,34]. However, when mice with deficient autophagy in the liver were depleted for p62 a dramatic protein aggregation phenotype was reversed [35]. These apparently opposing results suggested that p62, in addition to mediating degradation, is also required for aggregate formation. The fact that polymerization of p62 via the PB1 domain is required for aggregate formation and efficient autophagic degradation, supports this hypothesis [4]. Formation of large protein aggregates is regarded as a cellular defence mechanism, as the occurrence of many small sub-microscopic protein aggregates can be more damaging than fewer and larger aggregates [36]. The aggresome, formed in response to proteasomal inhibition or overexpression of aggregation prone proteins, is currently the best-studied protein aggregate with respect to formation and degradation mechanisms [30,37]. However, in this case it is likely not to be the single large aggresome localized at the microtubule-organizing centre of the cell that is degraded by autophagy, but rather smaller aggregates within or disentangled from the aggresome.

The Hsp70 (heat-shock protein 70) complex mediates quality control of newly synthesized proteins in the cytosol. Functional proteins are released or delivered to the Hsp90 chaperone complex, which monitors the condition of mature proteins. Hsp90 protects proteins from unfolding and aggregation, whereas Hsp70 is responsible for their degradation in cases when unfolding or aggregation cannot be prevented. In order to be degraded by the UPS, a cargo must be polyubiquitinated with chains consisting of four or more preferably Lys^{48}-linked ubiquitin moieties. If the capacity of the proteasome is overwhelmed, ubiquitin chains on misfolded proteins can undergo remodelling by the combined activity of DUBs (de-ubiquitinating enzymes) and E3 ligases to remove and add ubiquitin chains of different linkages (Figure 5). The newly formed ubiquitin chains, which contain Lys^{63}-linked chains, are recognized by p62 and HDAC6 (histone

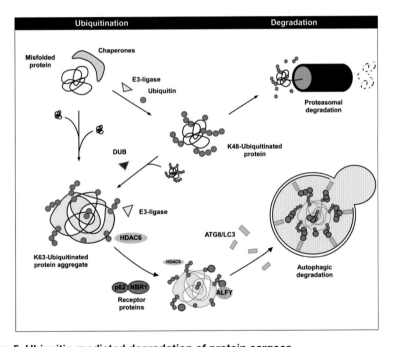

Figure 5. Ubiquitin-mediated degradation of protein cargoes
Misfolded proteins and protein aggregates that the chaperone complexes are unable to refold or untangle are labelled with ubiquitin and degraded by the proteasome or by aggrephagy. Single polypeptides tagged with Lys^{48}-linked ubiquitin chains are directed to the proteasome for degradation. Protein aggregates may be labelled by different ubiquitin chains, but Lys^{63}-linked chains may be more important in recruiting HDAC6 and autophagy receptors, such as p62 and NBR1, for subsequent degradation by selective autophagy. HDAC6 is involved in both the transport of Lys^{63}–ubiquitin-labelled aggregates and in the fusion of autophagosomes with lysosomes. ALFY exits from the nucleus to assist p62 in aggregate formation and subsequent autophagic degradation. DUBs and E3 ligases are involved in the tagging and editing of the proteins and protein aggregates destined for degradation.

© 2013 Biochemical Society

deacetylase 6), which direct protein aggregates to the aggresome and autophagy. HDAC6 facilitates dynein-mediated transport of ubiquitinated cargoes to the aggresome, and it is also important for the clearance of aggresomes by autophagy [30]. In aggrephagy involving p62 and NBR1, the large adaptor protein ALFY (autophagy-linked FYVE) is also involved in facilitating aggregate formation and subsequent autophagic degradation of these aggregates [38] (Figure 5).

Xenophagy

Pathogenic bacteria enter the cell through an invagination of the outer membrane, creating a phagosome targeted for lysosomal degradation in a process called LAP (LC3-associated phagocytosis). Following internalization, some bacteria escape the phagosome to proliferate and spread to neighbouring cells. The association of ubiquitin with cytosolic bacteria led to the hypothesis that antimicrobial autophagy requires an ubiquitin-dependent mechanism for cargo recognition and degradation, analogous to aggrephagy. Three of the SLRs, p62, NDP52 and optineurin, have been found to recognize ubiquitinated bacteria and facilitate sequestering to autophagosomes [19,20,39]. Depletion of either protein causes increased replication of *Salmonella typhimurium* and all three proteins are independently recruited to the same bacterium. To escape the phagosome, bacteria excrete peptides that damage the vacuolar membrane. NDP52 is recruited to *Salmonella* in damaged vacuoles through its binding to cytosolic galectin-8. The latter binds host vacuolar glycans exposed to the cytosol upon vacuole damage by bacteria. Since galectin-8 also signals sterile damage to endosomes or lysosomes, it serves as a versatile reporter of vesicle damage recruiting NDP52. After initial recruitment of NDP52, ubiquitination occurs, recruiting more NDP52 and the other SLRs for selective autophagic removal of damaged vesicles and pathogens [40]. Perhaps the most important role of autophagy receptors during xenophagy is to reduce inflammation by degrading membrane remnants and limit inflammatory cytokine signalling. The professional intracellular pathogens have evolved mechanisms to exploit and interfere with the host's xenophagy response. For example, some bacteria survive and replicate in the phagosomes by excreting DUBs to counter recognition by the SLRs, whereas, i.e. *Listeria monocytogenes*, manipulate the host actin-nucleation machinery to promote intracellular motility and intercellular spreading. In this process the bacteria surround themselves with cellular proteins to escape host cell recognition [41].

Regulation of selective autophagy

The autophagy pathway is directly regulated by several kinases, including ULK1/2 (uncoordinated 51-like kinase 1/2) and mTORC1 [mammalian (also known as mechanistic) target of rapamycin complex 1]. The autophagy receptors p62 and optineurin, as well as LC3B, have been shown recently to be regulated by phosphorylation. TBK1 [TANK (tumour-necrosis-factor-receptor-associated factor-associated nuclear factor κB activator)-binding kinase 1] phosphorylates the UBA domain of p62 to increase its binding to ubiquitinated cargoes and boost their autophagic degradation [42,43]. TBK1 is also a key regulator of immunological autophagy and is responsible for the maturation of autophagosomes into lytic bactericidal organelles [43]. Optineurin is phosphorylated in its LIR motif by TBK1 to dramatically increase the affinity for binding to ATG8 family proteins, again to increase selective autophagy [20]. LC3B has also been shown to be negatively regulated by phosphorylation in the N-terminal arm by protein kinase A [44].

Little is known about differential gene regulation occurring as a result of aggregate formation, but there is clearly a link between aggregate formation and oxidative stress responses. KEAP1 [Kelch-like ECH (erythroid cell-derived protein with cap 'n' collar homology)-associated protein 1] is an E3 ligase that mediates continuous degradation of NFR2 (NF-E2-related factor 2), a transcription factor responsible for activation of genes involved in the oxidative stress response. In a positive-feedback loop, p62 interacts with KEAP1 to disrupt the degradation of NRF2, thereby activating the oxidative stress response. Expression of p62 is itself under the control of NRF2, resulting in an amplification of the oxidative stress response as long as p62 levels rise [2].

In addition to regulation by phosphorylation, it is anticipated that other post-translational modifications including acetylation and ubiquitination will be implicated in the regulation of selective autophagy.

Concluding remarks

A distinction is usually made between basal housekeeping autophagy, important in quality control of proteins and organelles, and starvation- or stress-induced autophagy. In higher eukaryotes it seems likely that all basal autophagy is contributed by selective autophagy, whereas the starvation-induced and part of the stress-induced autophagy is more regularly associated with bulk degradation. However, our increased understanding of this process undoubtedly raises the question of the degree of selectivity involved. One could imagine that as a consequence of multi-cellularity autophagy has evolved to become increasingly more selective. A significant future challenge will be to define the spatiotemporal mechanisms by which ubiquitin and other signals control selective autophagy and the importance of selective autophagy in physiological and pathophysiological processes. Assessing the role of selective autophagy in cellular well-being will be vitally important to meet the acute challenges of age-related diseases that become more frequent as populations grow increasingly older.

Summary
- Selective autophagy plays a significant role in macroautophagy.
- In selective autophagy the forming autophagosome is enriched for specific cargoes, including organelles, protein aggregates, midbody rings and invading pathogens.
- Selective autophagy is mediated by autophagy receptor proteins that interact with cargo and the ATG8/LC3 family proteins.
- Autophagy receptors interact with ATG8 family proteins via LIR motifs, and ubiquitin is often involved in the labelling and recognition of cargo.
- Selective autophagy is regulated by signalling pathways mediating phosphorylation of several of the autophagy receptors.

We apologize to those whose work we were unable to include due to space constraints. This work was supported by grants from the Biology and Biomedicine (FRIBIO) program of the Norwegian Research Council and the Norwegian Cancer Society (to T.J.).

References

1. Nakatogawa, H., Suzuki, K., Kamada, Y. and Ohsumi, Y. (2009) Dynamics and diversity in autophagy mechanisms: lessons from yeast. Nat. Rev. Mol. Cell Biol. **10**, 458–467
2. Johansen, T. and Lamark, T. (2011) Selective autophagy mediated by autophagic adapter proteins. Autophagy **7**, 279–296
3. Lynch-Day, M.A. and Klionsky, D.J. (2010) The Cvt pathway as a model for selective autophagy. FEBS Lett. **584**, 1359–1366
4. Bjørkøy, G., Lamark, T., Brech, A., Outzen, H., Perander, M., Øvervatn, A., Stenmark, H. and Johansen, T. (2005) p62/SQSTM1 forms protein aggregates degraded by autophagy and has a protective effect on huntingtin-induced cell death. J. Cell Biol. **171**, 603–614
5. Pankiv, S., Clausen, T.H., Lamark, T., Brech, A., Bruun, J.A., Outzen, H., Overvatn, A., Bjorkoy, G. and Johansen, T. (2007) p62/SQSTM1 binds directly to Atg8/LC3 to facilitate degradation of ubiquitinated protein aggregates by autophagy. J. Biol. Chem. **282**, 24131–24145
6. Moscat, J., Diaz-Meco, M.T. and Wooten, M.W. (2007) Signal integration and diversification through the p62 scaffold protein. Trends Biochem. Sci. **32**, 95–100
7. Zatloukal, K., Stumptner, C., Fuchsbichler, A., Heid, H., Schnoelzer, M., Kenner, L., Kleinert, R., Prinz, M., Aguzzi, A. and Denk, H. (2002) p62 Is a common component of cytoplasmic inclusions in protein aggregation diseases. Am. J. Pathol. **160**, 255–263
8. Birgisdottir, A.B., Lamark, T. and Johansen, T. (2013) The LIR motif: crucial for selective autophagy. J. Cell Sci. **126**, 3237–3247
9. Deretic, V. (2012) Autophagy as an innate immunity paradigm: expanding the scope and repertoire of pattern recognition receptors. Curr. Opin. Immunol. **24**, 21–31
10. Shaid, S., Brandts, C.H., Serve, H. and Dikic, I. (2013) Ubiquitination and selective autophagy. Cell Death Differ. **20**, 21–30
11. Hanna, R.A., Quinsay, M.N., Orogo, A.M., Giang, K., Rikka, S. and Gustafsson, A.B. (2012) Microtubule-associated protein 1 light chain 3 (LC3) interacts with Bnip3 protein to selectively remove endoplasmic reticulum and mitochondria via autophagy. J. Biol. Chem. **287**, 19094–19104
12. Kanki, T., Wang, K., Cao, Y., Baba, M. and Klionsky, D.J. (2009) Atg32 is a mitochondrial protein that confers selectivity during mitophagy. Dev. Cell **17**, 98–109
13. Liu, L., Feng, D., Chen, G., Chen, M., Zheng, Q., Song, P., Ma, Q., Zhu, C., Wang, R., Qi, W. et al. (2012) Mitochondrial outer-membrane protein FUNDC1 mediates hypoxia-induced mitophagy in mammalian cells. Nat. Cell Biol. **14**, 177–185
14. Novak, I., Kirkin, V., McEwan, D.G., Zhang, J., Wild, P., Rozenknop, A., Rogov, V., Lohr, F., Popovic, D., Occhipinti, A. et al. (2010) Nix is a selective autophagy receptor for mitochondrial clearance. EMBO Rep. **11**, 45–51
15. Okamoto, K., Kondo-Okamoto, N. and Ohsumi, Y. (2009) Mitochondria-anchored receptor Atg32 mediates degradation of mitochondria via selective autophagy. Dev. Cell **17**, 87–97
16. Jiang, S., Wells, C.D. and Roach, P.J. (2011) Starch-binding domain-containing protein 1 (Stbd1) and glycogen metabolism: identification of the Atg8 family interacting motif (AIM) in Stbd1 required for interaction with GABARAPL1. Biochem. Biophys. Res. Commun. **413**, 420–425
17. Suzuki, K., Kondo, C., Morimoto, M. and Ohsumi, Y. (2010) Selective transport of alpha-mannosidase by autophagic pathways: identification of a novel receptor, Atg34p. J. Biol. Chem. **285**, 30019–30025
18. Kirkin, V., Lamark, T., Sou, Y.S., Bjorkoy, G., Nunn, J.L., Bruun, J.A., Shvets, E., McEwan, D.G., Clausen, T.H., Wild, P. et al. (2009) A role for NBR1 in autophagosomal degradation of ubiquitinated substrates. Mol. Cell **33**, 505–516
19. Thurston, T.L., Ryzhakov, G., Bloor, S., von Muhlinen, N. and Randow, F. (2009) The TBK1 adaptor and autophagy receptor NDP52 restricts the proliferation of ubiquitin-coated bacteria. Nat. Immunol. **10**, 1215–1221

20. Wild, P., Farhan, H., McEwan, D.G., Wagner, S., Rogov, V.V., Brady, N.R., Richter, B., Korac, J., Waidmann, O., Choudhary, C. et al. (2011) Phosphorylation of the autophagy receptor optineurin restricts *Salmonella* growth. Science **333**, 228–233
21. Ichimura, Y., Kumanomidou, T., Sou, Y.S., Mizushima, T., Ezaki, J., Ueno, T., Kominami, E., Yamane, T., Tanaka, K. and Komatsu, M. (2008) Structural basis for sorting mechanism of p62 in selective autophagy. J. Biol. Chem. **283**, 22847–22857
22. Deosaran, E., Larsen, K.B., Hua, R., Sargent, G., Wang, Y., Kim, S., Lamark, T., Jauregui, M., Law, K., Lippincott-Schwartz, J. et al. (2013) NBR1 acts as an autophagy receptor for peroxisomes, J. Cell Sci. **126**, 939–952
23. Svenning, S., Lamark, T., Krause, K. and Johansen, T. (2011) Plant NBR1 is a selective autophagy substrate and a functional hybrid of the mammalian autophagic adapters NBR1 and p62/SQSTM1. Autophagy **7**, 993–1010
24. Korac, J., Schaeffer, V., Kovacevic, I., Clement, A.M., Jungblut, B., Behl, C., Terzic, J. and Dikic, I. (2012) Ubiquitin-independent function of optineurin in autophagic clearance of protein aggregates. J. Cell Sci. **126**, 580–592
25. Gibbings, D., Mostowy, S., Jay, F., Schwab, Y., Cossart, P. and Voinnet, O. (2012) Selective autophagy degrades DICER and AGO2 and regulates miRNA activity. Nat. Cell Biol. **14**, 1314–1321
26. Noda, N.N., Kumeta, H., Nakatogawa, H., Satoo, K., Adachi, W., Ishii, J., Fujioka, Y., Ohsumi, Y. and Inagaki, F. (2008) Structural basis of target recognition by Atg8/LC3 during selective autophagy. Genes Cells **13**, 1211–1218
27. Behrends, C., Sowa, M.E., Gygi, S.P. and Harper, J.W. (2010) Network organization of the human autophagy system. Nature **466**, 68–76
28. von Muhlinen, N., Akutsu, M., Ravenhill, B.J., Foeglein, A., Bloor, S., Rutherford, T.J., Freund, S.M., Komander, D. and Randow, F. (2012) LC3C, bound selectively by a noncanonical LIR motif in NDP52, is required for antibacterial autophagy. Mol. Cell **48**, 329–342
29. van Wijk, S.J., Fiskin, E., Putyrski, M., Pampaloni, F., Hou, J., Wild, P., Kensche, T., Grecco, H.E., Bastiaens, P. and Dikic, I. (2012) Fluorescence-based sensors to monitor localization and functions of linear and K63-linked ubiquitin chains in cells. Mol. Cell **47**, 797–809
30. Lamark, T. and Johansen, T. (2012) Aggrephagy: selective disposal of protein aggregates by macroautophagy. Int. J. Cell Biol. **2012**, 736905
31. Tan, J.M., Wong, E.S., Kirkpatrick, D.S., Pletnikova, O., Ko, H.S., Tay, S.P., Ho, M.W., Troncoso, J., Gygi, S.P., Lee, M.K. et al. (2008) Lysine 63-linked ubiquitination promotes the formation and autophagic clearance of protein inclusions associated with neurodegenerative diseases. Hum. Mol. Genet. **17**, 431–439
32. Taylor, J.P., Hardy, J. and Fischbeck, K.H. (2002) Toxic proteins in neurodegenerative disease. Science **296**, 1991–1995
33. Hara, T., Nakamura, K., Matsui, M., Yamamoto, A., Nakahara, Y., Suzuki-Migishima, R., Yokoyama, M., Mishima, K., Saito, I., Okano, H. and Mizushima, N. (2006) Suppression of basal autophagy in neural cells causes neurodegenerative disease in mice. Nature **441**, 885–889
34. Komatsu, M., Waguri, S., Chiba, T., Murata, S., Iwata, J., Tanida, I., Ueno, T., Koike, M., Uchiyama, Y., Kominami, E. and Tanaka, K. (2006) Loss of autophagy in the central nervous system causes neurodegeneration in mice. Nature **441**, 880–884
35. Komatsu, M., Waguri, S., Koike, M., Sou, Y.S., Ueno, T., Hara, T., Mizushima, N., Iwata, J., Ezaki, J., Murata, S. et al. (2007) Homeostatic levels of p62 control cytoplasmic inclusion body formation in autophagy-deficient mice. Cell **131**, 1149–1163
36. Arrasate, M., Mitra, S., Schweitzer, E.S., Segal, M.R. and Finkbeiner, S. (2004) Inclusion body formation reduces levels of mutant huntingtin and the risk of neuronal death. Nature **431**, 805–810
37. Kopito, R.R. (2000) Aggresomes, inclusion bodies and protein aggregation. Trends Cell Biol. **10**, 524–530

38. Clausen, T.H., Lamark, T., Isakson, P., Finley, K., Larsen, K.B., Brech, A., Overvatn, A., Stenmark, H., Bjorkoy, G., Simonsen, A. and Johansen, T. (2010) p62/SQSTM1 and ALFY interact to facilitate the formation of p62 bodies/ALIS and their degradation by autophagy. Autophagy **6**, 330–344
39. Zheng, Y.T., Shahnazari, S., Brech, A., Lamark, T., Johansen, T. and Brumell, J.H. (2009) The adaptor protein p62/SQSTM1 targets invading bacteria to the autophagy pathway. J. Immunol. **183**, 5909–5916
40. Thurston, T.L., Wandel, M.P., von Muhlinen, N., Foeglein, A. and Randow, F. (2012) Galectin 8 targets damaged vesicles for autophagy to defend cells against bacterial invasion. Nature **482**, 414–418
41. Mostowy, S. and Cossart, P. (2012) Bacterial autophagy: restriction or promotion of bacterial replication? Trends Cell Biol. **22**, 283–291
42. Matsumoto, G., Wada, K., Okuno, M., Kurosawa, M. and Nukina, N. (2011) Serine 403 phosphorylation of p62/SQSTM1 regulates selective autophagic clearance of ubiquitinated proteins. Mol. Cell **44**, 279–289
43. Pilli, M., Arko-Mensah, J., Ponpuak, M., Roberts, E., Master, S., Mandell, M.A., Dupont, N., Ornatowski, W., Jiang, S., Bradfute, S.B. et al. (2012) TBK-1 promotes autophagy-mediated antimicrobial defense by controlling autophagosome maturation. Immunity **37**, 223–234
44. Cherra, 3rd, S.J., Kulich, S.M., Uechi, G., Balasubramani, M., Mountzouris, J., Day, B.W. and Chu, C.T. (2010) Regulation of the autophagy protein LC3 by phosphorylation. J. Cell Biol. **190**, 533–539
45. Kim, P.K., Hailey, D.W., Mullen, R.T. and Lippincott-Schwartz, J. (2008) Ubiquitin signals autophagic degradation of cytosolic proteins and peroxisomes. Proc. Natl. Acad. Sci. U.S.A. **105**, 20567–20574
46. Farre, J.C., Manjithaya, R., Mathewson, R.D. and Subramani, S. (2008) PpAtg30 tags peroxisomes for turnover by selective autophagy. Dev. Cell **14**, 365–376
47. Motley, A.M., Nuttall, J.M. and Hettema, E.H. (2012) Pex3-anchored Atg36 tags peroxisomes for degradation in *Saccharomyces cerevisiae*. EMBO J. **31**, 2852–2868

Mitophagy

Thomas MacVicar[1]

School of Biochemistry, Medical Sciences Building, University of Bristol, University Walk, Bristol BS8 1TD, U.K.

Abstract

Mitophagy describes the selective targeting and degradation of mitochondria by the autophagy pathway. In this process, defective mitochondria are first purged from the mitochondrial network then delivered to the lysosome by the autophagy machinery. Mitophagy has emerged as a key facet of mitochondrial quality control and has been implicated in a variety of human diseases. Disturbances in the cellular control of mitophagy can result in a dysfunctional mitochondrial network with grave implications for high energy demanding tissue. The present chapter reviews the recent advancements in the study of mitophagy mechanisms and regulation.

Keywords:
Atg32, erythropoiesis, mitochondrial dynamics, mitochondrial dysfunction, Nix, parkin, Parkinson's disease, PINK1.

Introduction

For decades, mitochondria have been recognized as the powerhouses of eukaryotic cells and we now appreciate that they function in other key cellular processes such as calcium signalling and apoptosis. The mitochondrial network is an amazingly dynamic and adaptable organelle system that must remain healthy in order to meet changing demands for ATP. Indeed, disturbances in mitochondrial homoeostasis result in an increasingly damaged and dysfunctional mitochondrial network, leading to a range of human pathologies including ischaemia, diabetes and neurodegeneration [1]. Dysfunctional mitochondria have the potential to generate vast quantities of ROS (reactive oxygen species) and can become 'leaky', releasing pro-apoptotic proteins into the cytoplasm, often with catastrophic consequences. If unchecked, deleterious mtDNA (mitochondrial DNA) mutations are also likely to manifest within the network

[1] email tom.macvicar@bristol.ac.uk

and further compromise mitochondrial activity [2]. Cells must therefore efficiently remove and replace damaged mitochondria with healthy ones in order to maintain their energy supply.

Damaged mitochondria are targeted for degradation in the lysosomes via a specific autophagic pathway called mitophagy [3]. When mitophagy was first described, the distinction between it and the non-selective autophagy pathway was hazy. However, following the discovery of proteins that function solely in the autophagic degradation of mitochondria [4], it soon became clear that mitophagy has evolved as a unique quality control mechanism. By sequestering dysfunctional (or sometimes redundant) mitochondria into double membrane autophagosomes and trafficking them to lysosomes, cells are able to neutralize any threat posed by the faulty powerhouses. Over the last 10 years, developments in techniques such as live-cell imaging have allowed us to make exciting progress in the study of mitophagy dynamics and regulation.

Mitophagy mechanisms

Mitophagy has been studied in a number of different systems both *in vitro* and *in vivo* and it has emerged that mitochondrial degradation can be initiated by different environmental and developmental cues. It is now clear that mitophagy is not solely a defence mechanism triggered by mitochondrial damage. By exploring the molecular mechanisms of mitophagy in more detail, we can begin to understand how it has evolved as an essential feature of both cellular quality control and adaptation.

Mitophagy in yeast

Mitophagy in yeast depends on the delivery of mitochondria to the acidic lysosome-like vacuole for degradation. Yeast initiate mitophagy in response to a number of conditions including nitrogen starvation and rapamycin treatment [5]. They have proven to be important in demonstrating that mitophagy can proceed via tightly regulated mechanisms distinct from the common autophagy pathway. Proteins such as the OMM (outer mitochondrial membrane) protein Uth1p were first demonstrated to regulate mitophagy independently of autophagy under certain conditions [4]. Subsequent genetic screens identified another OMM protein, Atg32, as being essential for mitophagy [6,7]. In order to guide ill-fated mitochondria to the autophagosome, Atg32 binds directly with the resident autophagosome protein Atg8 and also interacts with the previously identified selective autophagy adaptor protein Atg11 [8] (Figure 1). Positive regulation of the Atg32–Atg11 interaction requires MAPK (mitogen-activated protein kinase)-dependent phosphorylation of Atg32 at Ser[114] [9]. This phosphorylation event is critical to mitophagy as mutation of the serine residue results in abolishment of the mitophagy response to nitrogen starvation [9]. Although Atg32 has no known higher eukaryote homologue, these seminal studies have certainly removed any doubts surrounding the selective nature of mitophagy.

Work in yeast has also advanced our understanding of how mitophagy is regulated during shifts in metabolism. For example, activation of mitophagy following an exchange of respiratory to fermentative conditions appears to signify a housekeeping role for mitophagy as redundant mitochondria are selected for degradation [5].

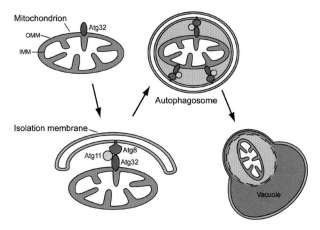

Figure 1. Mitophagy in yeast
Yeast mitochondria are targeted to the isolation membrane by the OMM protein Atg32. Atg32 can directly bind the autophagosome protein Atg8 or via interaction with an adapter protein Atg11. Atg11 may provide extra control of autophagosome assembly around mitochondria by interacting with both Atg32 and Atg8. Following the initial targeting, autophagosomes seal around mitochondria and eventually fuse with the acidic vacuole for degradation of the entrapped organelle [6–8].

Mitophagy during erythropoiesis

A striking example of mitophagy being employed to remove healthy yet redundant mitochondria occurs during mammalian red blood cell development (erythropoiesis). Mature red blood cells are devoid of mitochondria following the selective elimination of their entire mitochondrial population by a distinct mitophagy pathway [10]. During normal erythroid terminal differentiation, expression of the OMM BNIP3-like protein Nix is augmented [11] and this is necessary for complete mitochondrial depletion [12,13]. Importantly, an absence of Nix causes defective erythroid maturation and anaemia in mice [12].

Similar to Atg32 in yeast, Nix confers the selectivity for mitochondrial engulfment by the autophagosome. As discussed in Chapter 7 of this volume, Nix directly interacts with autophagosomal marker proteins LC3 (light-chain 3) and GABARAP (γ-aminobutyric acid receptor-associated protein) via its LIR (LC3-interacting region) motif [14] allowing the recruitment and formation of autophagosomes around the mitochondria (Figure 2). Interestingly, however, rescue of mitophagy in $Nix^{-/-}$ maturing blood cells is only partially suppressed by mutation of the LIR motif and recently a newly characterized 16-amino-acid cytoplasmic region of Nix called the MER (minimal essential region) has been described as critical for Nix activity [15]. Although LC3 does not interact with the MER [15], identification of other possible mitophagy-related interaction partners will improve our mechanistic understanding of mitochondrial clearance in developing red blood cells. Besides interacting directly with autophagosome components, Nix can also regulate mitophagy by facilitating mitochondrial membrane depolarization and ROS generation [16]. A Nix-regulated increase in ROS production may provide a key step in mitophagy by enhancing autophagosome biogenesis via mTOR [mammalian (also known as mechanistic) target of rapamycin] inhibition [16].

Nix-orchestrated mitophagy during erythropoiesis is arguably an example of pre-programmed mitochondrial clearance. How then do mammalian cells react to unexpected

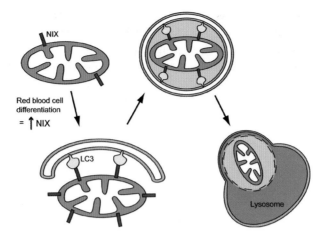

Figure 2. Nix regulated mitophagy during red blood cell development
Maturing red blood cells increase Nix expression on the OMM. Nix binds to LC3, the mammalian homologue of Atg8, through its cytoplasmic LIR motif. This mediates the engulfment of mitochondria by forming autophagosomes [11,12,15].

mitochondrial damage in order to preserve a healthy mitochondrial population? This question appears to be particularly significant in respiring differentiated cells such as neurons and hepatocytes. Given the pathological consequences of mitochondrial dysfunction in such tissue, extensive research has begun to elucidate the mechanisms that exist to selectively target and eradicate damaged mitochondria.

Parkin-mediated mitophagy and beyond

Mutations in the E3-ubiquitin ligase parkin [17] and PINK1 [PTEN (phosphatase and tensin homologue deleted on chromosome 10)-induced putative kinase 1] [18] were found to cause autosomal recessive forms of PD (Parkinson's disease); however, it was not initially clear how these two proteins contributed to the severe pathogenesis. The spotlight was cast on a possible role in mitochondrial quality control when *in vivo* studies demonstrated that these mutations manifest as mitochondrial dysfunction. *Drosophila* harbouring PINK1/parkin mutations suffer dopaminergic neuron loss and locomotion defects, associated with abnormal mitochondria [19,20]. Subsequently, elegant *Drosophila* genetics were employed to demonstrate that parkin functions downstream from PINK1 in the same pathway [20,21].

The question remained of how mutations in PINK1 and parkin might result in the accumulation of dysfunctional mitochondria. An answer was provided when Narendra et al. [22] discovered that parkin normally functions to target damaged mitochondria for degradation by mitophagy. Live-cell imaging of cultured cells expressing fluorescently tagged parkin revealed the remarkably dynamic recruitment of cytosolic parkin to damaged mitochondria. These parkin-decorated mitochondria are then sequestered into autophagosomes and delivered to lysosomes for degradation. Astoundingly, many parkin-overexpressing cells can clear their entire mitochondrial population after the induction of global mitochondrial damage by treatment with mitochondrial depolarizing drugs.

Questions remain of how physiologically relevant such wholesale examples of mitophagy are, not least because more bioenergetically active cells have been demonstrated to resist parkin-mediated mitophagy in response to depolarizing drugs [23]. In that study, cells with a greater dependence on oxidative phosphorylation were found to inhibit parkin translocation to depolarized mitochondria [23]. Important discrepancies also exist between the functions of endogenous and overexpressed parkin [24]. Further work examining the role of endogenous parkin, particularly in disease-vulnerable neurons, is essential. Surprisingly, parkin-knockout mice do not display hallmark PD phenotypes such as reduced motor ability and decreased neurological function [25]. It remains unclear why parkin-deficient fly and mouse models vary so much in the severity of Parkinsonism symptoms. Nevertheless, both *in vitro* and *in vivo* experiments continue to be essential for further dissection of the parkin/PINK1 mitophagy pathway.

In a healthy mitochondrion, PINK1 is rapidly turned over by constitutive proteolytic processing. However, following insult and dissipation of the inner membrane potential (mt$\Delta\Psi$), which is required to drive ATP production and mitochondrial protein import, PINK1 becomes stabilized on the OMM permitting it to recruit its binding partner parkin [26,27]. PINK1 interaction with parkin is necessary for parkin self-association and activation of parkin's ubiquitin ligase activity [28]. Once activated, parkin ubiquitinates targets on the OMM allowing the clustering of damaged mitochondria and their engulfment by autophagosomes (Figure 3).

Parkin is not the only ubiquitin ligase shown to direct mitophagy. Increased activity of the ubiquitin ligase Mul1 results in enhanced mitophagy during skeletal muscle wasting. Suppression of Mul1 maintains mitochondrial number and partially prevents muscle wasting in mice exposed to muscle-wasting stimuli [29]. Finally, gp78 (glycoprotein 78) is another E3 ubiquitin ligase recently identified to drive depolarization-induced mitophagy in a parkin-independent manner [30]. Interestingly, gp78 is ER (endoplasmic reticulum)-associated and,

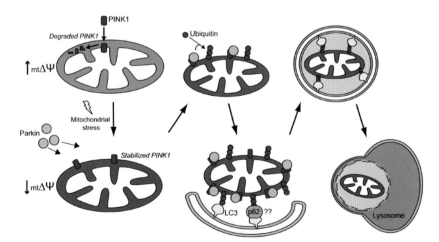

Figure 3. Parkin-mediated mitophagy
In a healthy mitochondrion with an active membrane potential (↑mt$\Delta\Psi$), PINK1 is imported to the IMM and constitutively turned over by proteolysis. However, following a compromise of mitochondrial activity and loss of membrane potential (↓mt$\Delta\Psi$), PINK1 stabilizes on the OMM allowing recruitment of the E3-ubiquitin ligase parkin. Parkin activation results in the ubiquitination of OMM proteins that act as an 'eat-me' signal recognized by a nascent autophagosome. LC3 can bind ubiquitin directly or may interact indirectly via adapter protein p62 [22,26].

© 2013 Biochemical Society

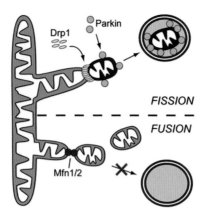

Figure 4. Mitochondrial dynamics and mitophagy
Mitochondrial fission is used to purge unwanted mitochondria from the network. The GTPase Drp1 drives dynamin-like scission of mitochondria. Parkin and other ubiquitin ligases can also facilitate the fission of depolarized mitochondria by targeting the mitofusins for degradation. Once divided from the network, exiled mitochondria can be engulfed by autophagosomes or, alternatively, if they remain healthy with an intact membrane potential, they may be permitted to re-fuse with the network. Mitochondrial fusion is coordinated by the OMM GTPases Mfn1 and Mfn2 and the IMM OPA1 (not shown). Enhanced fusion can protect mitochondria from autophagy degradation [45,46,49].

upon mitochondrial depolarization, its ubiquitination activity recruits LC3 to engulf mitochondria in close proximity to the ER [30].

Exactly how autophagosomes are recruited to parkin-decorated mitochondria remains an intriguing question. The adaptor protein p62 can bind both ubiquitin and LC3 and is found to accumulate on mitochondria that have been polyubiquitinated by parkin. Despite this, although a role for p62 in mitochondrial clustering looks likely [31], conflicting reports exist as to whether p62 is actually essential for the recruitment of autophagosomes to mitochondria (or vice versa) [31,32]. As we have seen with Atg32 in yeast and Nix in mammalian red blood cells, mitophagy does not always depend on cytosolic adapter proteins for recruitment of the autophagosome. Indeed, hypoxia-induced mitophagy also requires the direct interaction between an integral OMM protein FUNDC1 and autophagosomal LC3. FUNDC1 binds LC3 via its LIR motif and this interaction is enhanced on FUNDC1 dephosphorylation during hypoxic conditions [33]. Hypoxia-induced mitophagy therefore offers another mechanism whereby mitochondria are targeted for mitophagy by regulated and direct interaction between OMM and autophagosome proteins.

Regardless of targeting, further work is certainly required to elucidate the actual origins of the 'mitophagosome'. As discussed in Chapter 3 of this volume, several organelles have so far been implicated in autophagosome biogenesis including the mitochondria themselves [34]. It has also been suggested recently that the autophagic isolation membrane can form *de novo* on depolarized mitochondria as opposed to being recruited from elsewhere in the cytoplasm [35]. There is much left to learn about the role played by mitochondria in autophagosome biogenesis, however, it is now apparent that mitochondria can also regulate their own degradation by strict control of their morphology. The remainder of the present chapter focuses on the exciting progress being made in the understanding of how, and why, mitophagy is regulated by mitochondrial dynamics.

Mitophagy regulation by mitochondrial dynamics

It is impossible to discuss mitophagy without exploring the regulatory role played by mitochondrial dynamics. Mitochondrial fusion and fission have been implicated in many of the classical mitochondrial-associated cellular pathways including calcium signalling, apoptosis and the cell cycle. To explore how mitochondrial dynamics regulates mitophagy we must first introduce the fission and fusion machinery.

Mitochondrial dynamics: a balance between fusion and fission

Mitochondrial fission is driven by the dynamin-related protein Drp1 [36], and OMM proteins Fis1 [37] and Mff [38]. Unlike Fis1 and Mff, the GTPase Drp1 is a cytoplasmic protein that localizes to sites of mitochondrial fission in a rigorously regulated manner. Drp1 is the target of several key post-translational modifications including sumoylation, ubiquitination and phosphorylation [39]. These modifications either enhance or disrupt Drp1 activity in a number of ways such as by regulating Drp1 translocation to the mitochondria or by mediating Drp1 assembly into spiral structures at the future sites of mitochondrial division.

The mitochondrial fusion machinery also consists of three key proteins: the OMM GTPases Mfn1 (mitofusin 1) and Mfn2 [40] and the IMM (inner mitochondrial membrane) protein OPA1 (optic atrophy 1) [41]. Although structurally similar, Mfn1 and Mfn2 have been associated with separate functions. Of note, Mfn2 tethers mitochondria to the ER at contact sites that have been shown recently to play an important role in autophagosome biogenesis [42]. On the IMM, OPA1 is regulated by constitutive and inducible proteolytic cleavage and besides driving IMM fusion, OPA1 also serves to remodel cristae; key for ATP production and apoptosis [43].

Linking fission with mitophagy

It is generally accepted that mitochondrial fission is a pre-requisite for mitophagy in many mammalian cell types. As others have pointed out, it is logical for an elongated mitochondrion of over 5 µm in length to be chopped up before engulfment by 0.5-µm-diameter autophagosomes [44]. But beyond this, it is clear that the fission and fusion machinery offer invaluable quality control and can help determine whether a mitochondrion is destined for degradation. Twig et al. [45] first demonstrated elegantly that fission can generate fragmented mitochondria with a low membrane potential and reduced levels of OPA1. These mitochondria therefore have an inherent inability to re-fuse with the healthy mitochondrial network. Following their expulsion from the mitochondrial network, the mitochondria are finally removed by mitophagy.

In parkin-mediated mitophagy, parkin translocation to depolarized mitochondria triggers the ubiquitination and proteasome-dependent degradation of the mitofusins [46], thus further isolating dysfunctional mitochondria from the rest of the network. In other pathways, Mfn2 is targeted by the muscle-wasting mitophagy ubiquitin ligase Mul1 [29] and both Mfn1 and Mfn2 are ubiquitinated by the mitophagy regulator gp78 [30]. Therefore obstructing mitochondrial fusion by mitofusin degradation appears to be a common step in mitophagy initiation.

Mitochondrial parkin recruitment also causes degradation of the kinesin motor adaptor protein Miro [47]. This detaches kinesin from the mitochondrial surface leading to an arrest of mitochondrial motility that may facilitate mitophagy. Such a mechanism is particularly significant in neurons where mitochondrial transport is key for the distribution of functional mitochondria along axons to synapses. So by preventing the transport of damaged and potentially toxic mitochondria to these vital domains, PINK1 and parkin may be limiting the risk posed by dysfunctional mitochondria. In addition, Miro localizes to ER–mitochondrial contact sites and it remains to be seen whether disruption of ER–mitochondrial tethering is a key step in initiating mitophagy [48].

Fusion protects mitochondria from mitophagy

Given the importance of mitochondrial fission in assisting mitophagy, is it conceivable that mitochondria can resist mitophagy under certain conditions by inhibiting fission and promoting fusion? Owing to metabolic and bioenergetic demands, it sometimes becomes the case that mitophagy must be kept in check in order to maintain a respiring mitochondrial population. For example, mitophagy is blocked in yeast that are fed on a carbon source making mitochondria essential for metabolism [8]. An exciting possibility is that the metabolic status of a cell can feed into the mitochondrial dynamics machinery. Evidence for this comes from recent reports that describe how mitochondria undergo fusion and are spared from mitophagy during starvation and thus are able to maintain ATP levels. Cells can therefore protect their precious mitochondria during starvation-induced macroautophagy by maintaining an elongated mitochondrial network [49,50]. In both studies it was found that starvation-induced activation of PKA (protein kinase A) leads to inhibitory phosphorylation of fission factor Drp1 at Ser^{637} and dephosphorylation at Ser^{616}. This blocks the translocation of Drp1 to the mitochondria and hence inhibits mitochondrial fission.

Despite recent advances, there is a long way to go if we are to truly understand the crosstalk between mitochondrial dynamics and mitophagy and how this affects disease pathology. Mitochondrial elongation depends on the type of nutrient starvation indicating that tight control of mitochondrial dynamics depends on intricate signalling pathways yet to be properly defined [51]. It is also important to bear in mind that mitochondrial networks differ from cell type to cell type. In yeast, for example, mitophagy occurs independently of mitochondrial fission [52]. Interestingly, on simulation of fasting by nutrient deprivation in the presence of glucagon, mitophagy is actually enhanced in GFP–LC3-expressing primary mouse hepatocytes [53]. Morphologically, the mitochondria do not appear fused in these hepatocytes and autophagosomes are even found to form around mitochondria that have an intact membrane potential [53]. This provides strong evidence that mitochondrial depolarization is not always a pre-requisite for the selective removal of mitochondria.

Conclusion

Mitophagy is fundamental to mitochondrial quality control and homoeostasis and its importance is demonstrated by the pathological consequences of its mis-regulation. Since its initial recognition as a *bona fide* example of a specific autophagy pathway, the sheer diversity of mitophagy regulatory mechanisms continues to surprise. Cells regulate mitochondrial degradation not only through control of the autophagy machinery, but also via delicate tuning of mitochondrial fusion and fission. It remains to be seen whether other cellular processes linked with mitochondria also have a role to play in mitophagy regulation. Calcium, for instance, has emerged as a key modulator of autophagy [54]. Given the central role played by mitochondria in calcium signalling [55], it too could have important functions in mitophagy control.

Better understanding of the mechanisms involved in mitochondrial quality control will hopefully hold serious therapeutic potential for the number of devastating human diseases associated with mitochondrial dysfunction.

Summary

- Mitophagy is the selective degradation of dysfunctional or redundant mitochondria by the autophagy machinery.
- Disturbances in mitochondrial quality control have been implicated in a range of human pathologies including neurodegeneration.
- Mitophagy-specific proteins exist that target mitochondria for degradation by directly or indirectly interacting with autophagosomes, these include Atg32 in yeast and Nix in mammalian red blood cells.
- The PINK1/parkin pathway selectively targets damaged mitochondria and its disruption has been implicated in early-onset forms of PD.
- Mitochondrial dynamics play a key role in regulating mitophagy; fission isolates damaged mitochondria from the network, whereas fusion protects healthy mitochondria from engulfment by autophagosomes.

References

1. Chan, D.C. (2006) Mitochondria: dynamic organelles in disease, aging, and development. Cell **125**, 1241–1252
2. Youle, R.J. and Narendra, D.P. (2011) Mechanisms of mitophagy. Nat. Rev. Mol. Cell Biol. **12**, 9–14
3. Lemasters, J.J. (2005) Selective mitochondrial autophagy, or mitophagy, as a targeted defense against oxidative stress, mitochondrial dysfunction, and aging. Rejuvenation Res. **8**, 3–5
4. Kissova, I., Deffieu, M., Manon, S. and Camougrand, N. (2004) Uth1p is involved in the autophagic degradation of mitochondria. J. Biol. Chem. **279**, 39068–39074
5. Bhatia-Kissova, I. and Camougrand, N. (2010) Mitophagy in yeast: actors and physiological roles. FEMS Yeast Res. **10**, 1023–1034
6. Kanki, T., Wang, K., Cao, Y., Baba, M. and Klionsky, D.J. (2009) Atg32 is a mitochondrial protein that confers selectivity during mitophagy. Dev. Cell **17**, 98–109
7. Okamoto, K., Kondo-Okamoto, N. and Ohsumi, Y. (2009) Mitochondria-anchored receptor Atg32 mediates degradation of mitochondria via selective autophagy. Dev. Cell **17**, 87–97
8. Kanki, T. and Klionsky, D.J. (2008) Mitophagy in yeast occurs through a selective mechanism. J. Biol. Chem. **283**, 32386–32393
9. Aoki, Y., Kanki, T., Hirota, Y., Kurihara, Y., Saigusa, T., Uchiumi, T. and Kang, D. (2011) Phosphorylation of serine 114 on Atg32 mediates mitophagy. Mol. Biol. Cell **22**, 3206–3217

10. Kundu, M., Lindsten, T., Yang, C.Y., Wu, J., Zhao, F., Zhang, J., Selak, M.A., Ney, P.A. and Thompson, C.B. (2008) Ulk1 plays a critical role in the autophagic clearance of mitochondria and ribosomes during reticulocyte maturation. Blood **112**, 1493–1502
11. Aerbajinai, W., Giattina, M., Lee, Y.T., Raffeld, M. and Miller, J.L. (2003) The proapoptotic factor Nix is coexpressed with Bcl-xL during terminal erythroid differentiation. Blood **102**, 712–717
12. Sandoval, H., Thiagarajan, P., Dasgupta, S.K., Schumacher, A., Prchal, J.T., Chen, M. and Wang, J. (2008) Essential role for Nix in autophagic maturation of erythroid cells. Nature **454**, 232–235
13. Schweers, R.L., Zhang, J., Randall, M.S., Loyd, M.R., Li, W., Dorsey, F.C., Kundu, M., Opferman, J.T., Cleveland, J.L., Miller, J.L. and Ney, P.A. (2007) NIX is required for programmed mitochondrial clearance during reticulocyte maturation. Proc. Natl. Acad. Sci. U.S.A. **104**, 19500–19505
14. Novak, I., Kirkin, V., McEwan, D.G., Zhang, J., Wild, P., Rozenknop, A., Rogov, V., Lohr, F., Popovic, D., Occhipinti, A. et al. (2010) Nix is a selective autophagy receptor for mitochondrial clearance. EMBO Rep. **11**, 45–51
15. Zhang, J., Loyd, M.R., Randall, M.S., Waddell, M.B., Kriwacki, R.W. and Ney, P.A. (2012) A short linear motif in BNIP3L (NIX) mediates mitochondrial clearance in reticulocytes. Autophagy **8**, 1325–1332
16. Ding, W.X., Ni, H.M., Li, M., Liao, Y., Chen, X., Stolz, D.B., Dorn, 2nd, G.W. and Yin, X.M. (2010) Nix is critical to two distinct phases of mitophagy, reactive oxygen species-mediated autophagy induction and Parkin-ubiquitin-p62-mediated mitochondrial priming. J. Biol. Chem. **285**, 27879–27890
17. Kitada, T., Asakawa, S., Hattori, N., Matsumine, H., Yamamura, Y., Minoshima, S., Yokochi, M., Mizuno, Y. and Shimizu, N. (1998) Mutations in the parkin gene cause autosomal recessive juvenile parkinsonism. Nature **392**, 605–608
18. Valente, E.M., Abou-Sleiman, P.M., Caputo, V., Muqit, M.M., Harvey, K., Gispert, S., Ali, Z., Del Turco, D., Bentivoglio, A.R., Healy, D.G. et al. (2004) Hereditary early-onset Parkinson's disease caused by mutations in PINK1. Science **304**, 1158–1160
19. Greene, J.C., Whitworth, A.J., Kuo, I., Andrews, L.A., Feany, M.B. and Pallanck, L.J. (2003) Mitochondrial pathology and apoptotic muscle degeneration in *Drosophila* parkin mutants. Proc. Natl. Acad. Sci. U.S.A. **100**, 4078–4083
20. Park, J., Lee, S.B., Lee, S., Kim, Y., Song, S., Kim, S., Bae, E., Kim, J., Shong, M., Kim, J.M. and Chung, J. (2006) Mitochondrial dysfunction in *Drosophila* PINK1 mutants is complemented by parkin. Nature **441**, 1157–1161
21. Clark, I.E., Dodson, M.W., Jiang, C., Cao, J.H., Huh, J.R., Seol, J.H., Yoo, S.J., Hay, B.A. and Guo, M. (2006) *Drosophila* pink1 is required for mitochondrial function and interacts genetically with parkin. Nature **441**, 1162–1166
22. Narendra, D., Tanaka, A., Suen, D.F. and Youle, R.J. (2008) Parkin is recruited selectively to impaired mitochondria and promotes their autophagy. J. Cell Biol. **183**, 795–803
23. Van Laar, V.S., Arnold, B., Cassady, S.J., Chu, C.T., Burton, E.A. and Berman, S.B. (2011) Bioenergetics of neurons inhibit the translocation response of Parkin following rapid mitochondrial depolarization. Hum. Mol. Genet. **20**, 927–940
24. Rakovic, A., Shurkewitsch, K., Seibler, P., Grunewald, A., Zanon, A., Hagenah, J., Krainc, D. and Klein, C. (2013) Phosphatase and tensin homolog (PTEN)-induced putative kinase 1 (PINK1)-dependent ubiquitination of endogenous Parkin attenuates mitophagy: study in human primary fibroblasts and induced pluripotent stem cell-derived neurons. J. Biol. Chem. **288**, 2223–2237
25. Perez, F.A. and Palmiter, R.D. (2005) Parkin-deficient mice are not a robust model of parkinsonism. Proc. Natl. Acad. Sci. U.S.A. **102**, 2174–2179
26. Matsuda, N., Sato, S., Shiba, K., Okatsu, K., Saisho, K., Gautier, C.A., Sou, Y.S., Saiki, S., Kawajiri, S., Sato, F. et al. (2010) PINK1 stabilized by mitochondrial depolarization recruits Parkin to damaged mitochondria and activates latent Parkin for mitophagy. J. Cell Biol. **189**, 211–221

27. Narendra, D.P., Jin, S.M., Tanaka, A., Suen, D.F., Gautier, C.A., Shen, J., Cookson, M.R. and Youle, R.J. (2010) PINK1 is selectively stabilized on impaired mitochondria to activate Parkin. PLoS Biol. **8**, e1000298
28. Lazarou, M., Narendra, D.P., Jin, S.M., Tekle, E., Banerjee, S. and Youle, R.J. (2013) PINK1 drives Parkin self-association and HECT-like E3 activity upstream of mitochondrial binding. J. Cell Biol. **200**, 163–172
29. Lokireddy, S., Wijesoma, I.W., Teng, S., Bonala, S., Gluckman, P.D., McFarlane, C., Sharma, M. and Kambadur, R. (2012) The ubiquitin ligase Mul1 induces mitophagy in skeletal muscle in response to muscle-wasting stimuli. Cell Metab. **16**, 613–624
30. Fu, M., St Pierre, P., Shankar, J., Wang, P.T., Joshi, B. and Nabi, I.R. (2013) Regulation of mitophagy by the Gp78 E3 ubiquitin ligase. Mol. Biol. Cell **24**, 1153–1162
31. Narendra, D., Kane, L.A., Hauser, D.N., Fearnley, I.M. and Youle, R.J. (2010) p62/SQSTM1 is required for Parkin-induced mitochondrial clustering but not mitophagy; VDAC1 is dispensable for both. Autophagy **6**, 1090–1106
32. Geisler, S., Holmstrom, K.M., Skujat, D., Fiesel, F.C., Rothfuss, O.C., Kahle, P.J. and Springer, W. (2010) PINK1/Parkin-mediated mitophagy is dependent on VDAC1 and p62/SQSTM1. Nat. Cell Biol. **12**, 119–131
33. Liu, L., Feng, D., Chen, G., Chen, M., Zheng, Q., Song, P., Ma, Q., Zhu, C., Wang, R., Qi, W. et al. (2012) Mitochondrial outer-membrane protein FUNDC1 mediates hypoxia-induced mitophagy in mammalian cells. Nat. Cell Biol. **14**, 177–185
34. Hailey, D.W., Rambold, A.S., Satpute-Krishnan, P., Mitra, K., Sougrat, R., Kim, P.K. and Lippincott-Schwartz, J. (2010) Mitochondria supply membranes for autophagosome biogenesis during starvation. Cell **141**, 656–667
35. Itakura, E., Kishi-Itakura, C., Koyama-Honda, I. and Mizushima, N. (2012) Structures containing Atg9A and the ULK1 complex independently target depolarized mitochondria at initial stages of Parkin-mediated mitophagy. J. Cell Sci. **125**, 1488–1499
36. Smirnova, E., Griparic, L., Shurland, D.L. and van der Bliek, A.M. (2001) Dynamin-related protein Drp1 is required for mitochondrial division in mammalian cells. Mol. Biol. Cell **12**, 2245–2256
37. James, D.I., Parone, P.A., Mattenberger, Y. and Martinou, J.C. (2003) hFis1, a novel component of the mammalian mitochondrial fission machinery. J. Biol. Chem. **278**, 36373–36379
38. Otera, H., Wang, C., Cleland, M.M., Setoguchi, K., Yokota, S., Youle, R.J. and Mihara, K. (2010) Mff is an essential factor for mitochondrial recruitment of Drp1 during mitochondrial fission in mammalian cells. J. Cell Biol. **191**, 1141–1158
39. Chang, C.R. and Blackstone, C. (2010) Dynamic regulation of mitochondrial fission through modification of the dynamin-related protein Drp1. Ann. N.Y. Acad. Sci. **1201**, 34–39
40. Santel, A. and Fuller, M.T. (2001) Control of mitochondrial morphology by a human mitofusin. J. Cell Sci. **114**, 867–874
41. Misaka, T., Miyashita, T. and Kubo, Y. (2002) Primary structure of a dynamin-related mouse mitochondrial GTPase and its distribution in brain, subcellular localization, and effect on mitochondrial morphology. J. Biol. Chem. **277**, 15834–15842
42. Hamasaki, M., Furuta, N., Matsuda, A., Nezu, A., Yamamoto, A., Fujita, N., Oomori, H., Noda, T., Haraguchi, T., Hiraoka, Y. et al. (2013) Autophagosomes form at ER-mitochondria contact sites. Nature **495**, 389–393
43. Frezza, C., Cipolat, S., Martins de Brito, O., Micaroni, M., Beznoussenko, G.V., Rudka, T., Bartoli, D., Polishuck, R.S., Danial, N.N., De Strooper, B. and Scorrano, L. (2006) OPA1 controls apoptotic cristae remodeling independently from mitochondrial fusion. Cell **126**, 177–189
44. Gomes, L.C. and Scorrano, L. (2011) Mitochondrial elongation during autophagy: a stereotypical response to survive in difficult times. Autophagy **7**, 1251–1253
45. Twig, G., Elorza, A., Molina, A.J., Mohamed, H., Wikstrom, J.D., Walzer, G., Stiles, L., Haigh, S.E., Katz, S., Las, G. et al. (2008) Fission and selective fusion govern mitochondrial segregation and elimination by autophagy. EMBO J. **27**, 433–446

46. Tanaka, A., Cleland, M.M., Xu, S., Narendra, D.P., Suen, D.F., Karbowski, M. and Youle, R.J. (2010) Proteasome and p97 mediate mitophagy and degradation of mitofusins induced by Parkin. J. Cell Biol. **191**, 1367–1380
47. Wang, X., Winter, D., Ashrafi, G., Schlehe, J., Wong, Y.L., Selkoe, D., Rice, S., Steen, J., LaVoie, M.J. and Schwarz, T.L. (2011) PINK1 and Parkin target Miro for phosphorylation and degradation to arrest mitochondrial motility. Cell **147**, 893–906
48. Kornmann, B., Osman, C. and Walter, P. (2011) The conserved GTPase Gem1 regulates endoplasmic reticulum-mitochondria connections. Proc. Natl. Acad. Sci. U.S.A. **108**, 14151–14156
49. Gomes, L.C., Di Benedetto, G. and Scorrano, L. (2011) During autophagy mitochondria elongate, are spared from degradation and sustain cell viability. Nat. Cell Biol. **13**, 589–598
50. Rambold, A.S., Kostelecky, B., Elia, N. and Lippincott-Schwartz, J. (2011) Tubular network formation protects mitochondria from autophagosomal degradation during nutrient starvation. Proc. Natl. Acad. Sci. U.S.A. **108**, 10190–10195
51. Gomes, L.C., Di Benedetto, G. and Scorrano, L. (2011) Essential amino acids and glutamine regulate induction of mitochondrial elongation during autophagy. Cell Cycle **10**, 2635–2639
52. Mendl, N., Occhipinti, A., Muller, M., Wild, P., Dikic, I. and Reichert, A.S. (2011) Mitophagy in yeast is independent of mitochondrial fission and requires the stress response gene WHI2. J. Cell Sci. **124**, 1339–1350
53. Kim, I. and Lemasters, J.J. (2011) Mitochondrial degradation by autophagy (mitophagy) in GFP-LC3 transgenic hepatocytes during nutrient deprivation. Am. J. Physiol. Cell Physiol. **300**, C308–C317
54. Smaili, S.S., Pereira, G.J., Costa, M.M., Rocha, K.K., Rodrigues, L., do Carmo, L.G., Hirata, H. and Hsu, Y.T. (2013) The role of calcium stores in apoptosis and autophagy. Curr. Mol. Med. **13**, 252–265
55. Rizzuto, R., De Stefani, D., Raffaello, A. and Mammucari, C. (2012) Mitochondria as sensors and regulators of calcium signalling. Nat. Rev. Mol. Cell Biol. **13**, 566–578

Autophagy and cell death

Tohru Yonekawa and Andrew Thorburn[1]

Department of Pharmacology, University of Colorado Anschutz Medical Campus, Aurora, CO 80045, U.S.A.

Abstract

Autophagy is intimately associated with eukaryotic cell death and apoptosis. Indeed, in some cases the same proteins control both autophagy and apoptosis. Apoptotic signalling can regulate autophagy and conversely autophagy can regulate apoptosis (and most likely other cell death mechanisms). However, the molecular connections between autophagy and cell death are complicated and, in different contexts, autophagy may promote or inhibit cell death. Surprisingly, although we know that, at its core, autophagy involves degradation of sequestered cytoplasmic material, and therefore presumably must be mediating its effects on cell death by degrading something, in most cases we have little idea of what is being degraded to promote autophagy's pro- or anti-death activities. Because autophagy is known to play important roles in health and many diseases, it is critical to understand the mechanisms by which autophagy interacts with and affects the cell death machinery since this will perhaps allow new ways to prevent or treat disease. In the present chapter, we discuss the current state of understanding of these processes.

Keywords:
autophagic cell death, autophagy-mediated protection, caspase, Drosophila *baculovirus inhibitor of apoptosis repeat-containing ubiquitin-conjugating enzyme (dBRUCE), tumour-necrosis-factor-related apoptosis-inducing ligand (TRAIL).*

Introduction

Autophagy delivers cytoplasmic components to lysosomes for degradation and recycling of the degradation products, such as amino acids, carbohydrates and lipids that are used to synthesize new proteins and organelles or metabolized to supply energy. Of the three main types of

[1]To whom correspondence should be addressed (email andrew.thorburn@ucdenver.edu).

autophagy in mammalian cells, microautophagy, chaperone-mediated autophagy and macroautophagy, by far our best understanding of how autophagy regulates cell death relates to macroautophagy. In macroautophagy, which will simply be referred to as 'autophagy' from now on in the present chapter, a double-membrane structure, called the autophagosome, sequesters proteins, organelles and other cytoplasmic components then fuses with lysosomes for degradation. The central problem in understanding how autophagy regulates and is regulated by cell death mechanisms is therefore to understand how the sequestration and degradation of cellular components in autophagosomes and autolysosomes affects apoptosis and other death pathways and, conversely, how apoptosis may regulate autophagy. The term 'autophagy' is often used to mean both the formation of autophagosomes and the full process whereby the autophagosome's contents are subsequently degraded after fusion with the lysosome. It is important to remember that these two things are not necessarily the same. When we refer to 'autophagic flux', we mean that the whole process took place and something was degraded. However, one can imagine situations where the effect on the cell could be not due to the flux, but instead caused just by the formation of the autophagosomes themselves and, as will be discussed below, we now know that some mechanisms whereby cell death is regulated involve autophagosomes, but not necessarily autophagic flux, whereas others involve autophagosome formation and flux.

A considerable amount of research has been carried out in recent years in this area (for recent reviews see [1,2]). Much of this interest has been driven by the fact that manipulation of autophagy holds great promise for improving treatment of diverse diseases [3] with the central idea being that both diseases where we want to kill cells (e.g. cancer) and diseases where we want to improve cell survival (e.g. neurodegenerative disease) could benefit by manipulation of autophagy. However, as discussed below, the connections between autophagy and cell death are complicated. Thus before the practical application of autophagy manipulation can be undertaken to prevent or treat disease can be undertaken successfully, we need to understand how these complications will affect our plans.

An important concept that is often underappreciated is that, even when autophagy modulates the amount of cell death in a particular context, this need not indicate a direct mechanistic connection between the process of autophagy and the process of cell death. For example, it is well established that autophagy can protect against neurodegenerative disease [4]; indeed, neural deficiency in autophagy is sufficient to cause neurodegenerative disease [5,6]. Since neurodegenerative disease is ultimately caused by neurons dying, the obvious conclusion is that autophagy prevents neurons from dying. However, this does not necessarily mean that in this context the process of autophagy interacts with the process of apoptosis and directly inhibits it. Instead the simpler (and the evidence suggests more likely) explanation for this protection is that neurodegenerative disease is often caused by toxic protein aggregates, and since autophagy is a process that degrades proteins, autophagy protects neurons by ensuring that the toxic stimulus (i.e. the protein aggregates) are disposed of and never have a chance to activate cell death mechanisms. This appears to be the case in autophagy-deficient mice, where aggregates of ubiquitin-containing proteins are found in the autophagy-deficient neurons and it is these aggregates that are normally removed by autophagy that activate cell death pathways [5,6]. Similarly, autophagy can protect cells against nutrient and growth factor deprivation and under such circumstances autophagy inhibition leads to apoptosis [7,8]. However, this does not necessarily mean that autophagy was inhibiting apoptosis pathways in this case. It could simply mean that autophagic degradation of existing cellular macromolecules provides the nutrients and energy needed to maintain the essential functions required for cell survival during starvation, and that the cell therefore never activates the apoptosis pathway in the first place. In the present chapter, we focus not on cases like this where autophagy regulates cell death indirectly by altering the presence or absence of an upstream signal that eventually activates death pathways. Instead, we will focus on the questions such as, does the process of autophagy whereby certain substrates are degraded in the lysosome control the cell death machinery? And, how do the proteins that control autophagy interact with and regulate the cell death machinery?

Autophagy as a promoter of cell death

Mammalian cell death occurs by multiple mechanisms some of which are deliberate 'suicide' that are clearly programmed and some of which seem to be accidental or the result of damage [9]. Apoptosis is by far the best-described programmed death mechanism and it is now also clear that necrosis can be programmed in many instances. It has been proposed that 'autophagic cell death' is another type of 'active' programmed death [10]. The suggestion that autophagic cell death occurs came about on the basis of the frequent observation that cell death is often accompanied by high levels of autophagosomes and features of active autophagy. However, just because cells that will die induce autophagy before their demise, does not necessarily implicate a causal relationship between the autophagy and cell death. Indeed, high levels of autophagy could be an indication that the cell is attempting to survive by inducing autophagy and that it is only when this effort fails, that death occurs. Thus the question is: does the process of autophagy actually lead to the death of the cell? There are many publications in the literature concluding that death stimuli in mammalian cells and non-mammalian systems are caused by autophagy [11]; however, whether these situations are a case of autophagy being upstream of apoptosis or autophagy killing the cell by a mechanism that has nothing to do with the other established death mechanisms such as apoptosis is often unclear. So, does autophagy kill cells and, if so, how?

One situation where autophagy may cause cell death is in situations where the cell has no ability to activate canonical apoptosis, which is usually the preferred mechanism of death. One of the first such examples was described in cells from BAX and BAK double-knockout mice, which lack the two main proteins that regulate the release of mitochondrial proteins during apoptosis. These cells are severely compromised in apoptosis and do not display any activation of pro-apoptotic caspases, but still die in response to stimuli such as DNA damaging agents [12] (in fact most stressful stimuli that kill cells by inducing apoptosis generally still kill cells when apoptosis is prevented). This non-apoptotic death was dependent on autophagy regulators ATG5 (autophagy-related protein 5) and Beclin1 and knockdown of these genes was sufficient to provide long-term protection to the cells after treatment with death stimuli. This implies that, in the absence of apoptosis, the DNA damage activates autophagy that kills the cells. In another example, pharmacological inhibition of caspases was found to induce death of L929 fibroblasts and this too was reported to rely on autophagy regulators but not apoptosis [13] and to involve autophagic degradation of catalase that ultimately led to increased ROS (reactive oxygen species) that killed the cells [14]. It is unclear whether the explanation of catalase degradation being the ultimate cause of the death applies more broadly to other situations of autophagic cell death or just to the caspase inhibitor-treated L929 cells. However, the major concern with these examples is that they represent a very artificial situation. Normal cells are probably never as defective in apoptosis as BAX/BAK double-knockout murine embryonic fibroblasts. Thus even if autophagy is inducing cell death in these contexts, this may be merely a reflection that the cells use this pathway to die because they have no other option rather than an indication that autophagy is a *bona fide* death mechanism that is able to promote death in cells under more normal circumstances.

Forced expression of oncogenic Ras was reported recently to induce very high levels of autophagy that led to autophagy-dependent death and loss of clonogenic capacity in an ovarian cancer cell line [15]. Cell death occurring after proliferative arrest, was inhibited by pharmacological and genetic inhibition of autophagy and was not associated with any detectable caspase activation or other markers of apoptosis. However, unlike the examples above where apoptosis was artificially blocked, the dying cells, as described in the study by Elgendy et al. [15], were still able to activate canonical caspase-dependent apoptosis in response to other stimuli, i.e. they are not unable to activate caspase-dependent apoptosis, they just do not do so in this instance. Interestingly, death in this case required Ras-induced expression of NOXA, a Bcl family protein that is well known as a regulator of p53-induced apoptosis [16]. Examples like this one suggest that autophagy can induce cell death even in cells where apoptosis has not been artificially blocked. However, this example is also rather artificial; the death stimulus was a massive forced overexpression of a mutated signalling protein and one might therefore reasonably wonder if it is physiologically significant.

However, under certain circumstances, the process of autophagy does indeed seem to promote cell death under normal physiological situations. The best understood examples occur during the cell death events that drive tissue remodelling occur during *Drosophila* development [17–19]. However, it is important to note that even in these cases, it does not necessarily follow that autophagy alone kills the cells. Instead, these examples of autophagy-dependent cell death under physiological conditions are largely due to autophagy making other death pathways (especially caspase-dependent apoptosis) more likely. For example, during *Drosophila* oogenesis, autophagy controls developmental cell death by selectively degrading the protein dBRUCE (*Drosophila* baculovirus inhibitor of apoptosis repeat-containing ubiquitin-conjugating enzyme), which itself functions to inhibit caspase activation [19] (Figure 1). Thus, in this case, autophagy promotes cell death by degrading a particular target that itself acts as a negative regulator of caspase activity thus stimulating higher levels of caspase activity, i.e. something that we would normally classify as being a form of apoptosis [10]. Additionally, at least some developmental cell death in flies occurs in an autophagy-dependent but caspase-independent manner [20], although in this case the underlying target that is degraded by autophagy and which causes the caspase-independent death is unclear.

The dBRUCE example is a case where autophagic flux is necessary for the effect on cell death, that is, the dBRUCE protein was degraded and this is why apoptosis was stimulated. Sometimes, however, autophagosome formation is important but flux is not. An example comes from recent work where it was shown that autophagosomes induce apoptosis by serving as a platform on which caspases (specifically caspase-8) can be activated [21]. This mechanism does not necessarily require any actual degradation of autophagosome cargo and so may not indicate a need for autophagic flux in order to get induction of apoptosis and cell death. However, it does still require the core autophagy machinery that is required for autophagosome formation. This

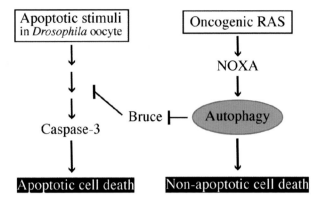

Figure 1. The promotion of cell death by autophagy
In *Drosophila* oocytes, apoptotic cell death occurs via caspase-3. Autophagy targets and degrades dBRUCE, an inhibitor of apoptosis, to promote apoptosis. Therefore inhibition of autophagy prevents nurse cells in late oocytes from causing apoptotic cell death. In some mammalian cells, overexpression of oncogenic mutated RAS increases autophagy level and leads to cell death that is dependent on autophagy but not apoptosis.

distinction has important practical implications. Suppose we wanted to block disease-associated apoptosis where autophagy is involved by a flux-dependent or a flux-independent mechanism; how would this affect the intervention in the autophagy process that is needed to achieve the goal of inhibiting apoptosis? In a case such as that in fly oocytes where autophagic flux and degradation of a substrate is required to alter sensitivity to apoptosis, it may be sufficient just to block the degradation step without affecting autophagosome formation in order to get the result you want. This can be carried out by inactivating lysosomes and we actually have drugs such as chloroquine that can do this. However, if the way that autophagy promotes apoptosis is because autophagosomes serve as a platform on which caspase activation is achieved, then just blocking the lysosome would have no effect. Instead, in this case, an intervention would need to prevent formation of the autophagosomes if you want to prevent apoptosis.

Thus although it seems clear that autophagy can promote cell death and the literature is filled with examples where claims of 'autophagic' death can be found, we still lack a detailed understanding of whether this can really occur under normal situations in a manner that is truly independent of other better established death mechanisms such as caspase-dependent apoptosis. And, in most cases, it is quite unclear how autophagy, i.e. a process that degrades things, is mediating its effects. The example of dBRUCE degradation [19] suggests that there may be other pro-survival signalling proteins that, if degraded, would lead to autophagy-dependent death. However, although it makes sense that they should exist, the full spectrum of such molecules is unknown. Without knowing what autophagy has to degrade in order to cause cell death, one cannot claim to understand mechanisms of autophagy-dependent death.

Autophagy as a protector against cell death

Although autophagy has been described as a death mechanism, the current consensus is that autophagy's role with regard to cell death is primarily protective [22,23]. However, as noted above, we need to be careful not to be confused by protective effects that, although real, are not necessarily indicative of a direct functional role of autophagy in blocking the death machinery. As explained above in the case of aggregate-prone proteins killing neurons, autophagy might protect without having any direct effect on the death pathways that actually carry out programmed cell deaths. With this in mind, how good is the evidence that autophagy and, in particular, autophagic flux to degrade cargo sequestered in autophagosomes directly inhibits apoptosis (or other death pathways)?

The first thing to consider is that autophagy has been reported to protect against a very diverse set of death stimuli. For example, a myriad of different chemotherapeutic agents used in cancer treatment have been reported to be sensitized by autophagy inhibition [24], implying that in each case autophagy is acting to reduce death in response to the chemotherapy in question. Since these agents work by very diverse means (e.g. DNA damage, interference with metabolic pathways, inhibition of steroid receptors, disruption of cytoskeleton, interference with growth-promoting kinase pathways, activation of cell-surface death receptor signalling and inhibition of the proteasome), it is hard to imagine that autophagy could always be merely working to remove a toxic stimulus as we discussed in the case of aggregated proteins in neurodegenerative disease. Instead, since the vast majority of these instances of autophagy protecting against cell death are in fact protection against apoptotic death, a simpler explanation would be that autophagy directly inhibits the apoptotic machinery.

Apoptosis is usually considered to occur by two distinct, but connected, pathways. The extrinsic pathway occurs after activation of death receptors such as Fas/CD95 or the TRAIL (tumour-necrosis-factor-related apoptosis-inducing ligand) receptors [25] and these pathways are well understood and have even been mathematically modelled [26]. Following ligand binding to the receptor, an adaptor protein called FADD (Fas-associated death domain) is recruited to the active receptor. FADD recruits the inactive pro-form of caspase-8 and this receptor–adaptor–caspase complex, which has been called DISC (death-inducing signalling complex), allows the dimerization and catalytic activation of caspase-8. Proteolytically active caspase-8 then cleaves itself and either directly cleaves the effector caspase, caspase-3, or more commonly cleaves a protein called Bid, which translocates to the mitochondria to cause activation of the intrinsic apoptosis pathway. Other apoptotic stimuli also use the intrinsic pathway, which is activated by pro-apoptotic BH3 (Bcl-2 homology domain 3)-only proteins from the Bcl family of apoptosis regulators (Bid is one of this family) that induce mitochondrial permeabilization causing release of cytochrome c and other mitochondrial proteins. Cytochrome c binds to a scaffolding protein called Apaf1 to form a framework called the apoptosome for caspase-9 activation (i.e. the apoptosome functions rather similarly to the DISC in that it creates a scaffold on which dimerization of the caspase that will start the ball rolling, caspase-8 for the DISC and caspase-9 for the apoptosome). In both pathways, small amounts of active caspase-3 (and caspase-7) are rapidly amplified through self-cleavage and cleavage of other caspases such as caspase-6, and the resulting burst of high-intensity caspase protease activity cleaves hundreds of proteins to result in the common morphology (e.g. cellular fragmentation, contraction and breakup of the nucleus) that is seen as apoptosis.

Given the very large number of examples where autophagy protects against apoptosis, it is somewhat surprising that we have few mechanistic examples which can explain how autophagy inhibits the apoptotic machinery. Thus, although hundreds of papers convincingly demonstrate that autophagy protects against apoptosis in response to very different stimuli, the key mechanistic question, what is it that autophagy is degrading to do this, is unknown for the vast majority of cases. One exception comes from examination of the effects of autophagy on TRAIL receptor signalling. As mentioned above, death receptors work in a quite straightforward way, the activated receptor recruits caspase-8 to the DISC and it is the catalytic activation of this caspase that initiates the cascade of proteolysis that leads to apoptosis. It has been reported that autophagy can inhibit TRAIL-induced apoptosis and at least in part and in certain cell lines, this appears to be due to autophagic degradation of cleaved caspase-8, but not the uncleaved (and inactive precursor) [27] (Figure 2). However, other mechanisms that would more generally apply to apoptosis whereby autophagy could inhibit the core apoptosis machinery have not yet been identified. One possibility is that since autophagy degrades mitochondria (a process known as mitophagy), this reduces the sensitivity of cells to apoptotic stimuli. However, since mitophagy occurs preferentially for impaired mitochondria with low membrane potential [28,29] and apoptosis involves the release of mitochondrial proteins from all or almost all the mitochondria in the cell, it is not immediately obvious why this on its own would lead to apoptosis inhibition by autophagy unless the mitochondrial mass was greatly reduced, which in most cases is not the case.

Thus a major open question in the field is to determine how autophagy has generally inhibitory effects on many apoptotic stimuli. One more radical possibility is that autophagy does not generally regulate the core apoptosis machinery, rather that autophagy regulators do so but in an autophagy-independent fashion. It is these activities that are responsible for the observed inhibitory effect of autophagy on apoptosis stimuli. The conclusion that autophagy protects against apoptosis is on the basis of the following type of experiment. An apoptotic stimulus (e.g. a chemotherapy drug) is shown to have increased effectiveness (i.e. it kills more cells at an intermediate dose) when autophagy is blocked either pharmacologically or genetically. By far the most common pharmacological inhibitors that have been used in the studies reported to date are agents that affect the lysosome, most commonly chloroquine. For genetic inhibition, the most common genes that are targeted are those encoding ATG proteins that are

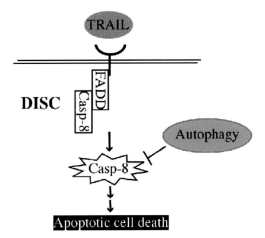

Figure 2. Example of the evasion from apoptotic cell death by autophagy
TRAIL induces DISC formation and caspase-8 is converted into active caspase-8 by self-processing in the DISC, which transduces a downstream signal cascade resulting in apoptosis. Autophagy can inhibit TRAIL-induced apoptosis by selective degradation of active caspase-8.

required for autophagosome formation [most often ATG5, ATG7, ATG12 or Beclin1, although sometimes other core ATGs such as LC3 (light-chain 3) are also targeted], and the inhibition is most commonly carried out by RNA interference. The rationale behind these experiments is that the inhibition of autophagy is specific, i.e. autophagy and only autophagy is inhibited by the inhibitor and if sensitivity to apoptosis is increased, this means that autophagy inhibited apoptosis. The problem is that these conclusions may be mistaken because the rationale is not correct, the manipulation may not truly be specific for autophagy. This is not too surprising with pharmacological inhibitors of autophagy, which are well known to be non-specific, e.g. chloroquine inhibits lysosomes and lysosomes do other things (including regulation of apoptosis) that may be unrelated to autophagy. Indeed, a recent paper demonstrated that chloroquine can sensitize cancer cells to anticancer drugs in an autophagy-independent manner [30].

In fact, the same thing is true for the *ATG* genes that regulate autophagy; the proteins they encode do not just regulate autophagy, they also regulate other processes, especially apoptosis (Figure 3). Thus, for example, ATG7 interacts with and controls p53 [31]. ATG12 can regulate apoptosis by two quite separate autophagy-independent mechanisms [32,33], and ATG5 can also regulate apoptosis independently of autophagy [34]. UVRAG (UV radiation resistance-associated gene) induces autophagosome formation as part of the Beclin1 complex [35], but also regulates BAX activation at the mitochondrial membrane to control the intrinsic apoptosis pathway [36]. Conversely, well-known apoptosis regulators also control autophagy. This is particularly evident for members of the Bcl family of proteins. Anti-apoptotic Bcl-2 and Bcl-xL proteins do not just regulate apoptosis; they also inhibit autophagy [37]. In addition, the core autophagy regulator Beclin1 itself possesses one of the domains, a BH3 domain, that defines the Bcl family [38]. Moreover, many BH3-only proteins that work in concert with Bcl-2, Bcl-xL and BAX and BAK to regulate mitochondrial permeability [39] also regulate autophagy. As noted above, NOXA can induce autophagy [15]. However, other BH3 proteins, BAD and BNIP3, as well as drugs that serve as BH3 mimetics can disrupt the interaction between Beclin1 and Bcl-xL and promote

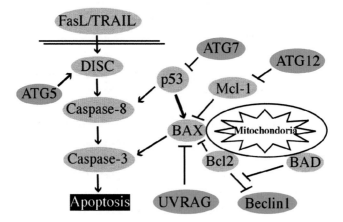

Figure 3. Cross-talk between autophagy and apoptosis regulators
Examples of autophagy regulators (blue) interacting with apoptosis regulators (grey). ATG5 helps activation of DISC via interaction with FADD. ATG7 and UVRAG can bind and inhibit apoptosis regulators, p53 and BAX respectively; at least theoretically this could explain a mechanism by which some autophagy proteins directly suppress apoptosis. ATG12 binds to anti-apoptotic members of the Bcl-2 family such as Mcl-1 and inhibits their anti-apoptotic activity. Conversely, Bcl-2 (and Bcl-xL) binds and inhibits an autophagy positive regulator Beclin1.

autophagy [40,41]. Conversely, another pro-apoptotic BH3 protein, BIM (Bcl-2-interacting mediator of cell death), can inhibit autophagy [42] and this is critical in the regulation of starvation-induced autophagy. Another well-known apoptosis regulator, FLIP, also controls autophagy [43]. The problem for our experiments which conclude that autophagy regulates apoptosis is clear. The experiment may have shown that ATG knockdown altered apoptosis, but this might have been due to an effect on an autophagy-independent function of that ATG protein. Additionally since many of these effects occur in different directions (a protein promotes death on one instance but inhibits in another), a problem might also arise where a true effect could be missed because of these competing effects. Imagine a case where an apoptotic stimulus kills 50% of the cells on a dish when ATG12 is knocked down. If autophagy protects against apoptosis, perhaps the amount of death that would have been obtained in the absence of knockdown would have been much less. However, since the ATG12 protein may also have been required for efficient induction of mitochondrial protein release through a completely different mechanism, e.g. inhibition of anti-apoptotic proteins like MCL-1 [33], the knockdown would at the same time inhibit this effect. Thus one could imagine a situation where the balance of pro- and anti-death effects of knocking down ATG12 may add up to little overall difference in the number of dead cells whether or not it was knocked down.

It is difficult to design experiments that discriminate between different effects by the same protein. One way is to identify mutant versions of the protein that affect one function but not the other. Debnath and colleagues recently used this approach of making such separation of function mutants to show that although ATG12 can be conjugated to ATG3, this does not affect autophagy as ATG12–ATG5 does, but rather affects mitochondrial homoeostasis and apoptosis [32]. The practical problem with this strategy is that mutants that really separate out different functions may be hard to identify; however, this strategy does allow one to at least begin to determine what function of a given autophagy regulator is important for the effects that are being measured.

In summary, we already have numerous examples where the same proteins, whether identified initially as an apoptosis regulator or as an autophagy regulator, also regulate the other process. This suggests a general model that regulation of both autophagy and apoptosis occurs again and again by the same proteins, but also raises a different potential explanation for the data suggesting that autophagy generally inhibits apoptosis; could some of these cases actually be due to the knockdown of autophagy regulators affecting steps in the apoptosis pathway without implicating any requirement for altered autophagic flux itself being necessary for the effects? Could the sequestration or depletion of a protein like UVRAG that otherwise might be involved in regulating the Beclin1 complex to control autophagosome formation result in less available UVRAG to inhibit BAX-induced apoptosis? Although such mechanisms still indicate that connections between autophagy and apoptosis are important, they lead to a very different view of how autophagy controls apoptosis, i.e. they suggest it may sometimes have nothing to do with the actual process of autophagosome formation, fusion with lysosomes and degradation of the contents.

Concluding remarks

We now have a large amount of evidence showing interactions between autophagy and cell death pathways, particularly the canonical apoptosis pathway. However, their larger significance for normal biology and disease is still unclear and we lack a full understanding of how mechanistically these connections take place. The first examples whereby autophagic degradation of specific targets such as dBRUCE or caspase-8 can promote or inhibit apoptosis have been found. However, in the vast majority of cases where autophagy appears to regulate apoptosis whether positively or negatively are still unclear in any mechanistic sense. Moreover, the big question, of how autophagy generally regulates the core apoptotic machinery, or even whether it does so, is unresolved. Given the intense focus on this area of research, better understanding of these mechanisms will certainly be obtained in the next few years. Furthermore, this knowledge may help guide the manipulation of these processes to treat diseases where we might want to see more or less cell death.

Summary

- There are some reports which demonstrate that autophagy promotes cell death. Apoptosis-null cells indicate a autophagy-dependent cell death in response to DNA damage agents and overexpression of oncogenic RAS causes apoptosis-independent cell death. Moreover, in *Drosophila* oocytes, autophagy is necessary for apoptotic cell death of nurse cells. Therefore autophagy can be needed for apoptotic and non-apoptotic cell death.
- Autophagy also inhibits apoptosis by degrading apoptosis regulators. For example, autophagy selectively degrades active caspase-8 in some TRAIL-induced apoptotic cells. Although we have many examples which show that autophagy protects cells against apoptosis, few reports explain those mechanisms.
- Recent studies have demonstrated that there are many physical and functional interactions between autophagy and apoptosis regulators. Some studies have shown that autophagy proteins regulate apoptosis proteins, whereas other studies have demonstrated that apoptosis proteins regulate autophagy proteins. Additionally, positive regulators of autophagy can positively or negatively (or differentially) regulate apoptosis proteins in different situations (or cases or contexts), and it is not clear whether this regulation is conditional or comprehensive. Further analysis is needed to solve these questions.

Work in our laboratory is supported by R01 grants from the National Institutes of Health [CA111421 and CA150925] and Shared Resources that are supported by P30 CA04934.

References

1. Gump, J.M. and Thorburn, A. (2011) Autophagy and apoptosis: what is the connection? Trends Cell Biol. **21**, 387–392
2. Rubinstein, A.D. and Kimchi, A. (2012) Life in the balance: a mechanistic view of the crosstalk between autophagy and apoptosis. J. Cell Sci. **125**, 5259–5268

© The Authors Journal compilation © 2013 Biochemical Society

3. Rubinsztein, D.C., Codogno, P. and Levine, B. (2012) Autophagy modulation as a potential therapeutic target for diverse diseases. Nat. Rev. Drug Discovery **11**, 709–730
4. Levine, B. and Kroemer, G. (2008) Autophagy in the pathogenesis of disease. Cell **132**, 27–42
5. Hara, T., Nakamura, K., Matsui, M., Yamamoto, A., Nakahara, Y., Suzuki-Migishima, R., Yokoyama, M., Mishima, K., Saito, I., Okano, H. and Mizushima, N. (2006) Suppression of basal autophagy in neural cells causes neurodegenerative disease in mice. Nature **441**, 885–889
6. Komatsu, M., Waguri, S., Chiba, T., Murata, S., Iwata, J., Tanida, I., Ueno, T., Koike, M., Uchiyama, Y., Kominami, E. and Tanaka, K. (2006) Loss of autophagy in the central nervous system causes neurodegeneration in mice. Nature **441**, 880–884
7. Lum, J.J., Bauer, D.E., Kong, M., Harris, M.H., Li, C., Lindsten, T. and Thompson, C.B. (2005) Growth factor regulation of autophagy and cell survival in the absence of apoptosis. Cell **120**, 237–248
8. Boya, P., Gonzalez-Polo, R.A., Casares, N., Perfettini, J.L., Dessen, P., Larochette, N., Metivier, D., Meley, D., Souquere, S., Yoshimori, T. et al. (2005) Inhibition of macroautophagy triggers apoptosis. Mol. Cell. Biol. **25**, 1025–1040
9. Green, D.R. and Victor, B. (2012) The pantheon of the fallen: why are there so many forms of cell death? Trends Cell Biol. **22**, 555–556
10. Galluzzi, L., Vitale, I., Abrams, J.M., Alnemri, E.S., Baehrecke, E.H., Blagosklonny, M.V., Dawson, T.M., Dawson, V.L., El-Deiry, W.S., Fulda, S. et al. (2012) Molecular definitions of cell death subroutines: recommendations of the Nomenclature Committee on Cell Death 2012. Cell Death Differ. **19**, 107–120
11. Denton, D., Nicolson, S. and Kumar, S. (2012) Cell death by autophagy: facts and apparent artefacts. Cell Death Differ. **19**, 87–95
12. Shimizu, S., Kanaseki, T., Mizushima, N., Mizuta, T., Arakawa-Kobayashi, S., Thompson, C.B. and Tsujimoto, Y. (2004) Role of Bcl-2 family proteins in a non-apoptotic programmed cell death dependent on autophagy genes. Nat. Cell Biol. **6**, 1221–1228
13. Yu, L., Alva, A., Su, H., Dutt, P., Freundt, E., Welsh, S., Baehrecke, E.H. and Lenardo, M.J. (2004) Regulation of an ATG7-Beclin 1 program of autophagic cell death by caspase-8. Science **304**, 1500–1502
14. Yu, L., Wan, F., Dutta, S., Welsh, S., Liu, Z., Freundt, E., Baehrecke, E.H. and Lenardo, M. (2006) Autophagic programmed cell death by selective catalase degradation. Proc. Natl. Acad. Sci. U.S.A. **103**, 4952–4957
15. Elgendy, M., Sheridan, C., Brumatti, G. and Martin, S.J. (2011) Oncogenic ras-induced expression of NOXA and Beclin-1 promotes autophagic cell death and limits clonogenic survival. Mol. Cell **42**, 23–35
16. Shibue, T., Takeda, K., Oda, E., Tanaka, H., Murasawa, H., Takaoka, A., Morishita, Y., Akira, S., Taniguchi, T. and Tanaka, N. (2003) Integral role of Noxa in p53-mediated apoptotic response. Genes Dev. **17**, 2233–2238
17. Berry, D.L. and Baehrecke, E.H. (2007) Growth arrest and autophagy are required for salivary gland cell degradation in *Drosophila*. Cell **131**, 1137–1148
18. Lee, C.Y., Cooksey, B.A. and Baehrecke, E.H. (2002) Steroid regulation of midgut cell death during *Drosophila* development. Dev. Biol. **250**, 101–111
19. Nezis, I.P., Shravage, B.V., Sagona, A.P., Lamark, T., Bjørkøy, G., Johansen, T., Rusten, T.E., Brech, A., Baehrecke, E.H. and Stenmark, H. (2010) Autophagic degradation of dBruce controls DNA fragmentation in nurse cells during late *Drosophila melanogaster* oogenesis. J. Cell Biol. **190**, 523–531
20. Denton, D., Shravage, B., Simin, R., Mills, K., Berry, D.L., Baehrecke, E.H. and Kumar, S. (2009) Autophagy, not apoptosis, is essential for midgut cell death in *Drosophila*. Curr. Biol. **19**, 1741–1746
21. Young, M.M., Takahashi, Y., Khan, O., Park, S., Hori, T., Yun, J., Sharma, A.K., Amin, S., Hu, C.-D., Zhang, J. et al. (2012) Autophagosomal membrane serves as platform for intracellular death-inducing signaling complex (iDISC)-mediated caspase-8 activation and apoptosis. J. Biol. Chem. **287**, 12455–12468

22. Levine, B. and Yuan, J. (2005) Autophagy in cell death: an innocent convict? J. Clin. Invest. **115**, 2679–2688
23. Kroemer, G. and Levine, B. (2008) Autophagic cell death: the story of a misnomer. Nat. Rev. Mol. Cell Biol. **9**, 1004–1010
24. Levy, J.M. and Thorburn, A. (2011) Targeting autophagy during cancer therapy to improve clinical outcomes. Pharmacol. Ther. **131**, 130–141
25. Thorburn, A. (2004) Death receptor-induced cell killing. Cell. Signalling **16**, 139–144
26. Albeck, J.G., Burke, J.M., Aldridge, B.B., Zhang, M., Lauffenburger, D.A. and Sorger, P.K. (2008) Quantitative analysis of pathways controlling extrinsic apoptosis in single cells. Mol. Cell **30**, 11–25
27. Hou, W., Han, J., Lu, C., Goldstein, L.A. and Rabinowich, H. (2010) Autophagic degradation of active caspase-8: a crosstalk mechanism between autophagy and apoptosis. Autophagy **6**, 891–900
28. Narendra, D., Tanaka, A., Suen, D.F. and Youle, R.J. (2008) Parkin is recruited selectively to impaired mitochondria and promotes their autophagy. J. Cell Biol. **183**, 795–803
29. Youle, R.J. and Narendra, D.P. (2011) Mechanisms of mitophagy. Nat. Rev. Mol. Cell Biol. **12**, 9–14
30. Maycotte, P., Aryal, S., Cummings, C.T., Thorburn, J., Morgan, M.J. and Thorburn, A. (2012) Chloroquine sensitizes breast cancer cells to chemotherapy independent of autophagy. Autophagy **8**, 200–212
31. Lee, I.H., Kawai, Y., Fergusson, M.M., Rovira, I.I., Bishop, A.J.R., Motoyama, N., Cao, L. and Finkel, T. (2012) Atg7 modulates p53 activity to regulate cell cycle and survival during metabolic stress. Science **336**, 225–228
32. Radoshevich, L., Murrow, L., Chen, N., Fernandez, E., Roy, S., Fung, C. and Debnath, J. (2010) ATG12 conjugation to ATG3 regulates mitochondrial homeostasis and cell death. Cell **142**, 590–600
33. Rubinstein, A.D., Eisenstein, M., Ber, Y., Bialik, S. and Kimchi, A. (2011) The autophagy protein Atg12 associates with antiapoptotic Bcl-2 family members to promote mitochondrial apoptosis. Mol. Cell **44**, 698–709
34. Yousefi, S., Perozzo, R., Schmid, I., Ziemiecki, A., Schaffner, T., Scapozza, L., Brunner, T. and Simon, H.U. (2006) Calpain-mediated cleavage of Atg5 switches autophagy to apoptosis. Nat. Cell Biol. **8**, 1124–1132
35. Liang, C., Feng, P., Ku, B., Dotan, I., Canaani, D., Oh, B.H. and Jung, J.U. (2006) Autophagic and tumour suppressor activity of a novel Beclin1-binding protein UVRAG. Nat. Cell Biol. **8**, 688–698
36. Yin, X., Cao, L., Kang, R., Yang, M., Wang, Z., Peng, Y., Tan, Y., Liu, L., Xie, M., Zhao, Y. et al. (2011) UV irradiation resistance-associated gene suppresses apoptosis by interfering with BAX activation. EMBO Rep. **12**, 727–734
37. Pattingre, S., Tassa, A., Qu, X., Garuti, R., Liang, X.H., Mizushima, N., Packer, M., Schneider, M.D. and Levine, B. (2005) Bcl-2 antiapoptotic proteins inhibit Beclin 1-dependent autophagy. Cell **122**, 927–939
38. Oberstein, A., Jeffrey, P.D. and Shi, Y. (2007) Crystal structure of the Bcl-XL-Beclin 1 peptide complex: Beclin 1 is a novel BH3-only protein. J. Biol. Chem. **282**, 13123–13132
39. Llambi, F., Moldoveanu, T., Tait, S.W.G., Bouchier-Hayes, L., Temirov, J., McCormick, L.L., Dillon, C.P. and Green, D.R. (2011) A unified model of mammalian BCL-2 protein family interactions at the mitochondria. Mol. Cell **44**, 517–531
40. Maiuri, M.C., Le Toumelin, G., Criollo, A., Rain, J.C., Gautier, F., Juin, P., Tasdemir, E., Pierron, G., Troulinaki, K., Tavernarakis, N. et al. (2007) Functional and physical interaction between Bcl-X(L) and a BH3-like domain in Beclin-1. EMBO J. **26**, 2527–2539
41. Zhang, H., Bosch-Marce, M., Shimoda, L.A., Tan, Y.S., Baek, J.H., Wesley, J.B., Gonzalez, F.J. and Semenza, G.L. (2008) Mitochondrial autophagy is an HIF-1-dependent adaptive metabolic response to hypoxia. J. Biol. Chem. **283**, 10892–10903

42. Luo, S., Garcia-Arencibia, M., Zhao, R., Puri, C., Toh, P.P.C., Sadiq, O. and Rubinsztein, D.C. (2012) Bim inhibits autophagy by recruiting Beclin 1 to microtubules. Mol. Cell **47**, 359–370
43. Lee, J.S., Li, Q., Lee, J.Y., Lee, S.H., Jeong, J.H., Lee, H.R., Chang, H., Zhou, F.C., Gao, S.J., Liang, C. and Jung, J.U. (2009) FLIP-mediated autophagy regulation in cell death control. Nat. Cell Biol. **11**, 1355–1362

10

Autophagy and ageing: implications for age-related neurodegenerative diseases

Bernadette Carroll, Graeme Hewitt and Viktor I. Korolchuk[1]

Institute for Ageing and Health, Newcastle University, Campus for Ageing and Vitality, Newcastle upon Tyne, NE4 5PL, U.K.

Abstract

Autophagy is a process of lysosome-dependent intracellular degradation that participates in the liberation of resources including amino acids and energy to maintain homoeostasis. Autophagy is particularly important in stress conditions such as nutrient starvation and any perturbation in the ability of the cell to activate or regulate autophagy can lead to cellular dysfunction and disease. An area of intense research interest is the role and indeed the fate of autophagy during cellular and organismal ageing. Age-related disorders are associated with increased cellular stress and assault including DNA damage, reduced energy availability, protein aggregation and accumulation of damaged organelles. A reduction in autophagy activity has been observed in a number of ageing models and its up-regulation via pharmacological and genetic methods can alleviate age-related pathologies. In particular, autophagy induction can enhance clearance of toxic intracellular waste associated with neurodegenerative diseases and has been comprehensively demonstrated to improve lifespan in yeast, worms, flies, rodents and primates. The situation, however, has been complicated by the identification that autophagy up-regulation can also occur during ageing. Indeed, in certain situations, reduced autophagosome induction may actually provide benefits to ageing cells. Future studies will undoubtedly improve our understanding of exactly how the multiple signals that are integrated to control appropriate autophagy activity change during ageing, what affect this has on autophagy and to what extent autophagy contributes to age-associated pathologies. Identification of mechanisms that influence a healthy lifespan is of economic, medical and social importance in our 'ageing' world.

[1] To whom correspondence should be addressed (email viktor.korolchuk@ncl.ac.uk).

Keywords:
AMPK, inflammation, mTOR, neurodegeneration, protein aggregation, ROS, cellular senescence.

Introduction
Three types of autophagy
Autophagy is characterized by the lysosomal degradation of substrates, however, the type of substrate as well as the mode of substrate recognition, transport and delivery differentiates the three known types of autophagy. First, macroautophagy is the term given to the isolation of cytoplasmic substrates into a double-membrane-bound organelle called an autophagosome. Autophagosomes are transported along the microtubule cytoskeleton network and their contents are degraded following fusion with the lysosome. A number of protein complexes have been demonstrated to be essential for macroautophagy induction. The ULK1 (uncoordinated-51-like kinase 1)-containing complex, the class III PI3K (phosphoinositide 3-kinase)–Beclin1 complex and the Atg5–Atg12–Atg16L1 (autophagy related 16-like 1) complex are indispensable for macroautophagy induction and participate in the initiation and growth of autophagosomes. Formation of an autophagosome also requires lipidation of LC3 (light-chain 3 or microtubule-associated protein 1 light chain 3; Atg8 in yeast) which is frequently used as a marker of macroautophagy. In addition, transport and fusion of autophagosomes with lysosomes is facilitated by a number of known proteins including Rab7 and SNAREs (soluble N-ethylmaleimide-sensitive fusion protein-attachment protein receptors) (Figure 1) [1]. Secondly, CMA (chaperone-mediated autophagy) utilizes the chaperone protein Hsc70 (heat-shock cognate 70 stress protein) to recognize specific soluble protein substrates. These substrates are delivered to the lysosomal membrane where they are unfolded and translocated into the lysosomal lumen via a Lamp2a-dependent mechanism (Figure 1) [2]. Thirdly, microautophagy is a process whereby cytoplasmic substrates are directly engulfed by lysosomes via invagination of the membrane (Figure 1) [3]. Little is known about how substrates are targeted for microautophagy and, furthermore, its activity during ageing has not been investigated to date. We will therefore limit the discussion in the present chapter to macroautophagy and CMA.

Upstream regulation of autophagy
The molecular mechanisms that regulate CMA are not well understood although increased activity has been observed in response to nutrient starvation, oxidative stress and hypoxia [2]. The mechanisms of macroautophagy regulation on the other hand have been studied in much greater detail. Nutrient starvation is a potent activator of autophagy and this occurs via the inhibition of the serine/threonine kinase mTOR [mammalian (also known as mechanistic) target of rapamycin] which occurs in complex with regulatory proteins [mTORC1 (mTOR complex 1)]. Integration of a number of mitogenic and growth-promoting signalling pathways converge on this core regulator, the best studied is the insulin–PI3K–Akt signalling axis [1].

Macroautophagy can also be activated in response to low-energy availability and is mediated by AMPK (AMP-activated protein kinase) and members of the Sirtuin family of deacetylases. Specifically, AMPK can induce macroautophagy via the inactivation of mTORC1 as well

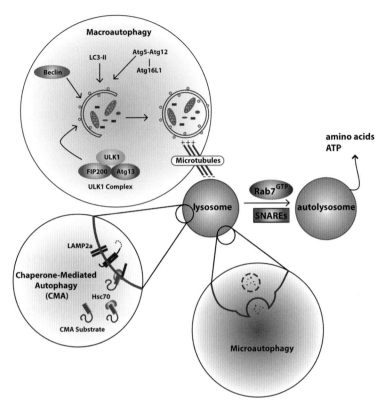

Figure 1. Overview of autophagy
Autophagy is a generic name for three major pathways of lysosome-dependent degradation of cytoplasmic components. Macroautophagy involves the sequestration of cytoplasmic contents including proteins and entire organelles into a double-membrane-bound autophagosome. Macroautophagy induction is regulated by a number of multiprotein complexes required for proper initiation, elongation and maturation of autophagosomes. These are transported along the microtubule network and fuse with lysosomes in a Rab7- and SNARE-dependent manner delivering their contents for hydrolytic degradation. CMA involves the recognition of protein substrates via a consensus peptide motif, KFERQ or similar, by the chaperone protein Hsc70. The protein substrates are delivered to the lysosome where they are unfolded and translocated across the lysosomal membrane into the acidic lumen via a Lamp2a-dependent mechanism. Microautophagy allows delivery of cytoplasmic contents to the lysosome lumen via direct invagination of the lysosome membrane.

as by direct phosphorylation of the autophagy-activating serine/threonine kinase ULK1 [4]. The most widely studied member of the Sirtuin family SIRT1 promotes macroautophagy by direct deacetylation of the Atg (autophagy-related) proteins Atg5, Atg12 and LC3 [5]. Both AMPK and SIRT1 can also activate the transcription factor FoxO3a which promotes the expression of a number of autophagy-related genes including LC3, Atg12, Bnip3 and Rab7 [6,7]. In addition, the effect of SIRT1 on macroautophagy can be mediated by the transcription factor p53 which can both induce (via transcriptional control of antioxidant Sestrin proteins that activate AMPK) and inhibit (when present in the cytoplasm) the process [6,8]. Other, mTOR-independent mechanisms regulating macroautophagy have also been identified, such as those involving free inositol levels in the cell [1,9].

Autophagy and ageing

Ageing is characterized by the accumulation of detrimental molecular, cellular and functional changes to an organism. It is associated with a gradual deterioration in cellular fitness and viability and an increase in cellular transformation. Age-related dysregulation of autophagy has been observed and, equally, dysregulated autophagy can contribute both directly and indirectly to cellular damage, such as an accumulation of damaged or toxic aggregates and organelles, thus compounding the problems associated with age. In the present chapter we discuss the current understanding of how ageing may cause a dysfunction in autophagy potential, including changes in transcription, translation and function of autophagy-related genes. We will also describe how modulation of autophagy can affect the ageing phenotype with a special focus on the role of autophagy in age-related neurodegenerative diseases.

Observed changes in autophagy during ageing

Altered expression and function of autophagy-related genes

Although ageing is often associated with a decrease in autophagy potential, current evidence does not paint a consistent picture as to the consequences of ageing on the expression of autophagy-related proteins (Figure 2). For example, both elevated [10] and decreased [11] expression of Beclin1 has been observed in aged tissues. Similar inconsistencies have also been observed in the expression of LC3, Atg5 and Atg7 [10–12]. In addition to altered expression of proteins, functional defects have also been observed during ageing. For example, age-related accumulation of lipofuscin can also significantly impair lysosomal function thereby reducing macroautophagy- and CMA-dependent substrate clearance. CMA is also impaired in aged rats, not as a result of decreased Lamp2a expression, but rather due to changes in the lysosomal membrane which causes a reduction in Lamp2a localization [13,14]. Appropriate levels of functional autophagy indeed contribute to maintaining health during ageing as demonstrated when Lamp2a levels are maintained into old age which preserves CMA levels in mice and ameliorates ageing-related phenotypes [13]. An area that has yet to be fully explored in ageing cells is the cross-talk and potential compensation mechanisms between the different modes of autophagy and the UPS (ubiquitin–proteasome system). Such cross-talk is important for the regulation of cellular homoeostasis. For example, macroautophagy is up-regulated in cells with defective CMA [15] and autophagy is up-regulated when the UPS system is impaired [16]. Specific cellular assaults and (dys)function during ageing are likely to contribute to the observed alterations in autophagy-related protein expression and function. A better understanding of the underlying mechanisms of ageing will undoubtedly help to elucidate the cause and effect of these changes in autophagy potential.

Reduced responsiveness to external and internal stimuli

The ability to sense and respond appropriately to external and internal stimuli is essential to maintain cellular homoeostasis. With increasing age, however, the responsiveness and activity of AMPK and SIRT1 are decreased thereby directly and indirectly contributing to the

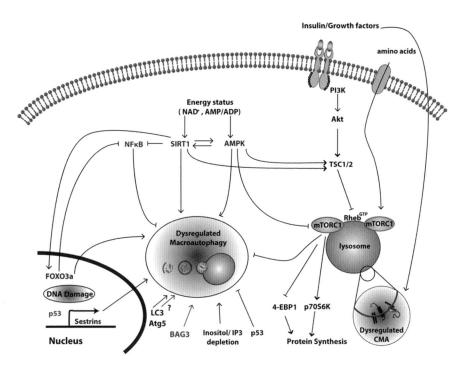

Figure 2. Signalling events involved in the regulation of autophagy
In the presence of anabolic signals only basal levels of autophagy occur in order to maintain homoeostasis. Autophagy is further activated in response to a number of stimuli including nutrient or energy deprivation. The regulation of autophagy occurs by transcription, translation and post-translational modifications of a multitude of proteins and complexes and these are affected by ageing. Proteins or pathways that show an age-dependent increase are denoted by red, whereas those with an identified age-dependent decrease are shown in blue. Ageing is often associated with a general decrease in cellular autophagy with an increase in inhibitory signals (such as NF-κB) and a decrease in sensitivity to stressors (reduced AMPK responsiveness); however, an increase in autophagy chaperone proteins such as BAG3 may indicate that complex environmental or genetic factors participate in the fate of autophagy during ageing. 4-EBP1, 4-E-binding protein 1; IP3, inositol 1,4,5-trisphosphate; TSC1/2, tuberous sclerosis complex 1/2.

dysregulation of autophagy (Figure 2). A number of studies have demonstrated that AMPK is not activated in old tissues under conditions where young tissues show a robust response [7]. The mechanisms of AMPK activity reduction are not well understood, but altered expression and activity of positive and negative regulators are likely to be important factors. SIRT1 activity may also decrease with age via decrease in expression [17] and decreased intracellular NAD$^+$ levels [18], leading to decreased cellular antioxidant capacity. Indeed, oxidative stress is particularly important during ageing and, in addition to SIRT1, other oxidative stress and autophagy regulators such as the polyamine spermidine are also reduced during ageing [19]. Reduction in FoxO3a [20] and p53 [7,21] activity have also been observed in ageing rodents which may also contribute to dysregulated autophagy (Figure 2).

Increased cellular damage and inflammatory responses

Low-grade chronic inflammation is a hallmark of ageing; however, there is a complex cause-and-effect relationship between inflammation, cellular damage and ageing, and the participation of autophagy in this process is under debate. The pro-inflammatory transcription factor NF-κB (nuclear factor-κB) is activated in response to cellular stress and DNA damage and its activity is increased in normal and accelerated ageing models [22]. Genetic or pharmacological inhibition of NF-κB can reduce ageing-related cell damage. The activity of the autophagy-inhibiting NF-κB is inversely correlated with the activity of the autophagy-activating AMPK and SIRT1 which, as discussed above, are reduced during ageing. Cross-talk between these signalling pathways and autophagy has yet to be fully investigated in ageing models. However, the careful balance between and maintenance of cellular stress responses is likely to be an important contributor to cellular ageing mechanisms. Indeed, increased age-related ROS (reactive oxygen species) for example, enhances cellular oxidative stress, inflammation, DNA damage and dysregulated mitochondria (Figure 3) [18,23,24]. Importantly, dysregulation of mitochondria causes a chronic reduction in cellular energy production, increased autophagy and reduced antioxidant capacity. This directly contributes to the inflammatory phenotype via further production of ROS [24] and ageing phenotypes. With regard to autophagy, on one hand, an increase in damaged mitochondria and ROS can promote autophagy up-regulation in an attempt to clear the damaged organelles. However, the extent to which autophagy can be activated is dependent on the expression and availability of its regulators, which as we have discussed can be altered during ageing. Furthermore, ROS production can also disrupt autophagy machinery, leading to a reduced capacity for autophagy clearance. Further work is required to elucidate the relationships between ROS, mitochondrial dysfunction and autophagy. It is clear that a wide range of complex interdependent relationships participate to tightly control cellular stress responses and maintain energy (Figures 2 and 3).

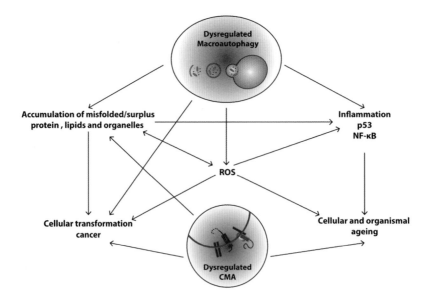

Figure 3. The general effect of dysregulated autophagy on cellular fate
Dysregulation of autophagy can have an impact on a variety of processes that ultimately affect cell health and survival. It is important to note that there are complex cause-and-effect relationships between many ageing-related cellular changes. For example, an accumulation of damaged mitochondria can contribute to increased intracellular ROS levels that can directly affect inflammatory responses as well as mitochondrial health and clearance. Such relationships can further compound age-related pathologies and have a direct impact on the regulation of autophagy.

Investigating the underlying mechanisms regulating inflammation, ROS, mitochondrial dysfunction and autophagy are important considerations for the future of ageing research.

Up-regulation of macroautophagy during ageing?

Reduced autophagy-related protein expression, function and responsiveness have all been suggested to contribute to cellular ageing. The situation, however, is not straightforward and in aged tissues an increase in macroautophagy has also been observed. Specifically, an increase in the expression of the co-chaperone protein BAG3, at the expense of BAG1, has been observed during ageing representing a shift from UPS- to macroautophagy-dependent degradation [25]. BAG3 can influence autophagy activity by interacting with the microtubule motor dynein and an adaptor protein called p62. It has been suggested that proteins destined for degradation are incorporated via BAG3 into p62-positive protein aggregates and targeted to autophagosomes [25,26]. Furthermore, BAG3 with a binding partner HspB8 can activate eIF2α and directly influence autophagy induction [27]. In addition, activation of autophagy has been observed in the mouse model of premature ageing, Hutchinson–Gilford progeria, and is associated with increased AMPK activity and decreased mTOR activity [28].

Additional work is required to further understand how autophagy regulation is altered during ageing. Specific environmental and genetic factors are likely to contribute to the precise modifications of autophagy-related protein expression and function during ageing. There is increasing evidence, however, that modulation of autophagy-related pathways can affect ageing and that carefully controlled autophagy could alleviate ageing-related pathologies and promote extension of a healthy lifespan.

The contribution of autophagy to ageing

Participation of autophagy in cellular senescence

Cellular senescence is often studied as a cell culture model of ageing although at an organismal level senescence can not only promote, but also prevent ageing by acting as a tumour suppression mechanism. Senescence refers to cells that have irreversibly exited the cell cycle (i.e. postmitotic), but remain metabolically active. It is considered to be a cell survival mechanism in the face of irreparable cellular damage that would otherwise lead to cellular transformation. Senescence can occur via replicative (telomere shortening) or non-replicative (DNA damage response or oncogene activation) mechanisms. Senescent cells are characterized by an increase in size and a SASP (senescence-associated secretory phenotype) mediated by secretion of cytokines and growth factors [29].

It is unclear at present what the true contribution of senescence to ageing is and indeed the extent to which autophagy contributes to senescence (Figure 3). For example, activation of autophagy has been shown to contribute to senescence in some studies of OIS (oncogene-induced senescence); mTORC1 activity is reduced, autophagy genes are up-regulated and genetic inhibition of autophagy delays the senescent phenotype [30]. However, in other studies, the mTORC1 inhibitor rapamycin has been shown to slow onset of senescence [31], possibly implicating reduced autophagy in senescence. Cellular senescence is an expanding field that in recent years has shifted focus from a molecular damage-induced (e.g. DNA damage and ROS production) senescence model to an mTORC1-centric model whereby mTORC1 stimulates ageing via hyper-activation of a cellular growth-promoting programme. The two models, however, are not necessarily mutually exclusive and future work in this area will certainly contribute important understanding to the role of mTOR and autophagy in senescence and the contribution of senescence to ageing.

Induction of autophagy contributes to increased lifespan

Studies designed to investigate factors that contribute to ageing have consistently identified proteins and pathways that modulate autophagy. One of the first mechanisms identified to increase lifespan was dietary restriction. The lifespan-promoting effect of this intervention appears to be evolutionarily conserved and has been observed in yeast, worms (*Caenorhabditis elegans*), flies (*Drosophila melanogaster*), rodents and primates. This increased longevity is mediated, at least in part, by the activation of autophagy, as lifespan extension is prevented by deletion or mutation of autophagy-related genes [32]. For example, mutation in the *C. elegans* Beclin1 homologue *bec-1* prevents the life-extending benefits of *daf-2* [insulin/IGF-1 (insulin-like growth factor-1) receptor] mutations [33]. Activators of autophagy including rapamycin, resveratrol and spermidine can also promote increased lifespan [19]. Cellular mTORC1 activity is important for longevity; it has been observed to persist during normal ageing and genetic or pharmacological down-regulation of the IGF-1–Akt–mTORC1 signalling axis can promote increased lifespan [34] and deletion of the downstream S6 protein kinase can extend lifespan of mice possibly via AMPK activation [35]. Modulation of upstream regulators of mTORC1 and autophagy provide further weight to the important contribution of autophagy to ageing. For example, activation of the autophagy-promoting proteins AMPK, SIRT1 and FoxO3a can also promote lifespan extension and inhibition of their activity can enhance ageing phenotypes [36], whereas loss of the pro-inflammatory p53 homologue in *C. elegans* can increase lifespan [37]. Specifically, SIRT1 may play a protective role to limit the detrimental effects of stressors such as ROS, as well as contributing to autophagy induction via autophagosome formation [6]. Indeed, lifespan extension mediated by dietary restriction is associated with increased SIRT1 expression and SIRT1 activators such as hypoxia can activate autophagy. The central importance of autophagy to promoting lifespan is further emphasized when one considers that different mechanisms of longevity, i.e. calorie restriction compared with rapamycin-induced are all ultimately dependent on autophagy-related genes.

Autophagy in age-related neurodegeneration

Autophagy is particularly important in post-mitotic or terminally differentiated cells such as neurons as they are unable to use 'mitotic dilution' as an option to remove toxic aggregates. Instead they rely on functional quality control mechanisms including autophagy and the UPS to remove potentially damaging proteins, complexes or entire organelles and are therefore extremely vulnerable when such quality control mechanisms go wrong. Parkinson's disease and Huntington's disease are associated with cytoplasmic aggregation of α-synuclein and mutant huntingtin respectively, and impaired function of autophagy pathways (Figure 4). For example, wild-type α-synuclein is ordinarily degraded by CMA and although Parkinson's disease-associated mutant α-synuclein can still be recruited to lysosomes, it is not translocated to the lysosome lumen. Furthermore, its association with CMA receptors subsequently impairs the ability of other substrates to bind and be degraded [38]. Genetic or

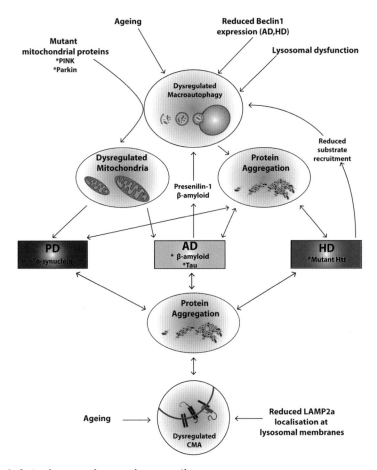

Figure 4. Autophagy and neurodegeneration
Dysregulation of autophagy in neurons has been implicated in neurodegenerative diseases including HD (Huntington's disease), AD (Alzheimer's disease) and PD (Parkinson's disease). The incidence of such diseases increases with age and many of the age-associated perturbations in autophagy regulation described in the present chapter are also associated with neurodegeneration. A particularly important relationship between ageing, neurodegeneration and autophagy is protein aggregation and its downstream consequences. For example, mutant tau proteins are observed in AD to cause neurofibrillar tangles thus perturbing membrane trafficking and it can block CMA machinery and thereby reduce overall CMA. In HD, expansion of the N-terminus of Htt (huntingtin) by multiple glutamine residues results in its accumulation and dysregulation of autophagy via multiple mechanisms. For example, autophagy perturbation can be caused by sequestration of Beclin1 into Htt aggregates; mutant Htt has also been shown to prevent recruitment of substrates to autophagic vesicles. In PD, mutant α-synuclein is recruited to CMA receptors, however, it is not translocated into the lysosome lumen. As a result, CMA receptors are unable to bind any further substrates and its activity is impaired. Dysfunctional mitochondria are associated with PD and AD. Mutations in the autophagy-associated mitochondrial proteins PINK and parkin prevent the clearance of damaged mitochondria in PD. Reduced expression (e.g. Beclin1 in AD and HD) and localization (Lamp2a) of proteins can perturb autophagy in ageing and neurodegenerative diseases. In AD cases caused by presenilin mutations, an increase in the number of cytoplasmic autophagosomes has been observed, possibly as the result of reduced lysosome acidification and therefore function. The accumulation of autophagosomes further contributes to AD pathology as a source of amyloid-β, which accumulates as a result of cleavage of the APP (amyloid precursor protein) and is secreted into the extracellular space where it forms amyloid plaques, a hallmark of AD.

pharmacological induction of autophagy can help to alleviate symptoms of neurodegenerative disorders by promoting degradation of mutant proteins in various animal models and increase lifespan [39].

Dysfunctional autophagy in neurodegenerative diseases can result from reduced expression and function of autophagy proteins (Figure 4). For example, a reduction in *Beclin1* expression has been noted in patients suffering from Alzheimer's disease and Huntington's disease. Beclin1 can also be sequestered into mutant huntingtin aggregates and is thus unable to induce autophagy. Increasing the expression of *Beclin1* allows restoration of functional autophagy and clearance of toxic proteins thus improving cellular pathology [38]. Mutations causing neurodegenerative diseases can also directly contribute to the impairment of autophagy. For example one of the consequences of Huntington's disease-causing polyglutamine expansion in huntingtin protein is defective autophagic substrate recruitment to autophagosomes. Indeed, a knock-in of mutant huntingtin lacking this polyglutamine tail, improves disease symptoms, promotes autophagosome formation and extends lifespan [40]. Furthermore, mutations of the ubiquitin ligase parkin and the serine/threonine kinase PINK1 [PTEN (phosphatase and tensin homologue deleted on chromosome 10)-induced putative kinase 1] are observed in sufferers of Parkinson's disease. These proteins are both autophagy-related proteins that participate in the specific targeting of damaged mitochondria for autophagic degradation and their mutation is associated with reduced autophagic clearance of mitochondria, increased mitochondrial dysfunction and oxidative stress thereby directly contributing to disease pathology [32,39,41].

In addition to protein aggregation and mitochondrial dysfunction, dysregulated autophagy in neurodegenerative diseases has also been observed as a result of reduced lysosome dysfunction. In Alzheimer's disease, a reduction in lysosomal lumen acidification has been associated with an accumulation of autophagosomes [42]. This is particularly important as autophagosomes have been implicated in the accumulation and subsequent extracellular deposition of amyloid-β which directly contributes to the disease pathology. Alzheimer's disease is also associated with dysfunctional hyper-phosphorylated tau proteins which ordinarily regulate microtubule dynamics, but in disease states contribute to neurofibrillar tangles and impaired membrane trafficking. Indeed neurons are particularly sensitive to impaired membrane trafficking events due to their elongated shape and defects in the microtubule network can have negative effects on the rate of autophagosome clearance. In such situations, improvements in lysosomal clearance or even inhibition of autophagosome synthesis have been suggested to be more beneficial than up-regulation of autophagy [43].

Conclusion

Dysregulated autophagy has been observed in ageing and age-related diseases and is an important mediator of pathology. Since the first demonstrations of a potential protective role for autophagy during ageing, many contributory and antagonistic regulatory mechanisms have been discovered. The potential to exploit autophagy as a therapeutic target for age-related pathologies including neurodegeneration is an attractive prospect. The challenge still remains, however, to identify medical interventions that promote appropriate levels of autophagy to maintain cellular homoeostasis.

Summary

- Dysregulation of autophagy has been observed in ageing tissues and is associated with altered expression and function of regulatory proteins and signalling pathways.
- Although many studies suggest autophagy is reduced during ageing, it must be noted that up-regulation of autophagy activity has also been demonstrated during ageing. It is likely that many factors contribute to the tight spatial and temporal activity of autophagy.
- Pharmacological and genetic inhibition of the insulin–Akt–mTORC1 signalling pathway is associated with increased lifespan via an evolutionarily conserved mechanism from yeast, to worms, flies, rodents and primates. These effects are abrogated in the absence of autophagy highlighting the central importance of this catabolic process in the promotion of lifespan.
- Neurodegenerative disorders such as Alzheimer's, Huntington's and Parkinson's diseases are often characterized by accumulation of protein aggregates. Dysregulated autophagy has been implicated in these pathologies and activation (less frequently suppression) of autophagy has been proposed as a promising therapeutic strategy.

Work in the laboratory is supported by an Early Career Award to V.I.K. from the Biotechnology and Biological Sciences Research Council (BBSRC). G.H. is supported by a Doctoral Training Grant studentship from the BBSRC.

References

1. Ravikumar, B., Sarkar, S., Davies, J.E., Futter, M., Garcia-Arencibia, M., Green-Thompson, Z.W., Jimenez-Sanchez, M., Korolchuk, V.I., Lichtenberg, M., Luo, S. et al. (2010) Regulation of mammalian autophagy in physiology and pathophysiology. Physiol. Rev. **90**, 1383–1435
2. Kaushik, S. and Cuervo, A.M. (2012) Chaperone-mediated autophagy: a unique way to enter the lysosome world. Trends Cell Biol. **22**, 407–417
3. Li, W.W., Li, J. and Bao, J.K. (2012) Microautophagy: lesser-known self-eating. Cell. Mol. Life Sci. **69**, 1125–1136
4. Inoki, K., Kim, J. and Guan, K.L. (2012) AMPK and mTOR in cellular energy homeostasis and drug targets. Annu. Rev. Pharmacol. Toxicol. **52**, 381–400
5. Lee, I.H., Cao, L., Mostoslavsky, R., Lombard, D.B., Liu, J., Bruns, N.E., Tsokos, M., Alt, F.W. and Finkel, T. (2008) A role for the NAD-dependent deacetylase Sirt1 in the regulation of autophagy. Proc. Natl. Acad. Sci. U.S.A. **105**, 3374–3379
6. Salminen, A. and Kaarniranta, K. (2009) SIRT1: regulation of longevity via autophagy. Cell. Signalling **21**, 1356–1360
7. Salminen, A. and Kaarniranta, K. (2012) AMP-activated protein kinase (AMPK) controls the aging process via an integrated signaling network. Ageing Res. Rev. **11**, 230–241
8. Tasdemir, E., Chiara Maiuri, M., Morselli, E., Criollo, A., D'Amelio, M., Djavaheri-Mergny, M., Cecconi, F., Tavernarakis, N. and Kroemer, G. (2008) A dual role of p53 in the control of autophagy. Autophagy **4**, 810–814

9. Sarkar, S., Floto, R.A., Berger, Z., Imarisio, S., Cordenier, A., Pasco, M., Cook, L.J. and Rubinsztein, D.C. (2005) Lithium induces autophagy by inhibiting inositol monophosphatase. J. Cell Biol. **170**, 1101–1111
10. Wohlgemuth, S.E., Seo, A.Y., Marzetti, E., Lees, H.A. and Leeuwenburgh, C. (2010) Skeletal muscle autophagy and apoptosis during aging: effects of calorie restriction and life-long exercise. Exp. Gerontol. **45**, 138–148
11. Lipinski, M.M., Zheng, B., Lu, T., Yan, Z., Py, B.F., Ng, A., Xavier, R.J., Li, C., Yankner, B.A., Scherzer, C.R. and Yuan, J. (2010) Genome-wide analysis reveals mechanisms modulating autophagy in normal brain aging and in Alzheimer's disease. Proc. Natl. Acad. Sci. U.S.A. **107**, 14164–14169
12. Simonsen, A., Cumming, R.C., Brech, A., Isakson, P., Schubert, D.R. and Finley, K.D. (2008) Promoting basal levels of autophagy in the nervous system enhances longevity and oxidant resistance in adult *Drosophila*. Autophagy **4**, 176–184
13. Kiffin, R., Kaushik, S., Zeng, M., Bandyopadhyay, U., Zhang, C., Massey, A.C., Martinez-Vicente, M. and Cuervo, A.M. (2007) Altered dynamics of the lysosomal receptor for chaperone-mediated autophagy with age. J. Cell Sci. **120**, 782–791
14. Zhang, C. and Cuervo, A.M. (2008) Restoration of chaperone-mediated autophagy in aging liver improves cellular maintenance and hepatic function. Nat. Med. **14**, 959–965
15. Kaushik, S., Massey, A.C., Mizushima, N. and Cuervo, A.M. (2008) Constitutive activation of chaperone-mediated autophagy in cells with impaired macroautophagy. Mol. Biol. Cell **19**, 2179–2192
16. Korolchuk, V.I., Menzies, F.M. and Rubinsztein, D.C. (2010) Mechanisms of cross-talk between the ubiquitin–proteasome and autophagy–lysosome systems. FEBS Lett. **584**, 1393–1398
17. Zheng, T. and Lu, Y. (2011) Changes in SIRT1 expression and its downstream pathways in age-related cataract in humans. Curr. Eye Res. **36**, 449–455
18. Braidy, N., Guillemin, G.J., Mansour, H., Chan-Ling, T., Poljak, A. and Grant, R. (2011) Age related changes in NAD$^+$ metabolism oxidative stress and Sirt1 activity in wistar rats. PLoS ONE **6**, e19194
19. Eisenberg, T., Knauer, H., Schauer, A., Buttner, S., Ruckenstuhl, C., Carmona-Gutierrez, D., Ring, J., Schroeder, S., Magnes, C., Antonacci, L. et al. (2009) Induction of autophagy by spermidine promotes longevity. Nat. Cell Biol. **11**, 1305–1314
20. Li, M., Chiu, J.F., Mossman, B.T. and Fukagawa, N.K. (2006) Down-regulation of manganese-superoxide dismutase through phosphorylation of FOXO3a by Akt in explanted vascular smooth muscle cells from old rats. J. Biol. Chem. **281**, 40429–40439
21. Feng, Z., Hu, W., Teresky, A.K., Hernando, E., Cordon-Cardo, C. and Levine, A.J. (2007) Declining p53 function in the aging process: a possible mechanism for the increased tumor incidence in older populations. Proc. Natl. Acad. Sci. U.S.A. **104**, 16633–16638
22. Tilstra, J.S., Clauson, C.L., Niedernhofer, L.J. and Robbins, P.D. (2011) NF-κB in aging and disease. Aging Dis. **2**, 449–465
23. Jenny, N.S. (2012) Inflammation in aging: cause, effect, or both? Discovery Med. **13**, 451–460
24. Sohal, R.S. and Orr, W.C. (2012) The redox stress hypothesis of aging. Free Radical Biol. Med. **52**, 539–555
25. Gamerdinger, M., Hajieva, P., Kaya, A.M., Wolfrum, U., Hartl, F.U. and Behl, C. (2009) Protein quality control during aging involves recruitment of the macroautophagy pathway by BAG3. EMBO J. **28**, 889–901
26. Gamerdinger, M., Kaya, A.M., Wolfrum, U., Clement, A.M. and Behl, C. (2011) BAG3 mediates chaperone-based aggresome-targeting and selective autophagy of misfolded proteins. EMBO Rep. **12**, 149–156
27. Carra, S. (2009) The stress-inducible HspB8–Bag3 complex induces the eIF2α kinase pathway: implications for protein quality control and viral factory degradation? Autophagy **5**, 428–429

28. Marino, G., Ugalde, A.P., Salvador-Montoliu, N., Varela, I., Quiros, P.M., Cadinanos, J., van der Pluijm, I., Freije, J.M. and Lopez-Otin, C. (2008) Premature aging in mice activates a systemic metabolic response involving autophagy induction. Hum. Mol. Genet. **17**, 2196–2211
29. Campisi, J. and d'Adda di Fagagna, F. (2007) Cellular senescence: when bad things happen to good cells. Nat. Rev. Mol. Cell Biol. **8**, 729–740
30. Young, A.R., Narita, M., Ferreira, M., Kirschner, K., Sadaie, M., Darot, J.F., Tavare, S., Arakawa, S., Shimizu, S., Watt, F.M. and Narita, M. (2009) Autophagy mediates the mitotic senescence transition. Genes Dev. **23**, 798–803
31. Demidenko, Z.N., Zubova, S.G., Bukreeva, E.I., Pospelov, V.A., Pospelova, T.V. and Blagosklonny, M.V. (2009) Rapamycin decelerates cellular senescence. Cell Cycle **8**, 1888–1895
32. Rubinsztein, D.C., Marino, G. and Kroemer, G. (2011) Autophagy and aging. Cell **146**, 682–695
33. Melendez, A., Talloczy, Z., Seaman, M., Eskelinen, E.L., Hall, D.H. and Levine, B. (2003) Autophagy genes are essential for dauer development and life-span extension in *C. elegans*. Science **301**, 1387–1391
34. Bjedov, I. and Partridge, L. (2011) A longer and healthier life with TOR down-regulation: genetics and drugs. Biochem. Soc. Trans. **39**, 460–465
35. Selman, C., Tullet, J.M., Wieser, D., Irvine, E., Lingard, S.J., Choudhury, A.I., Claret, M., Al-Qassab, H., Carmignac, D., Ramadani, F. et al. (2009) Ribosomal protein S6 kinase 1 signaling regulates mammalian life span. Science **326**, 140–144
36. Salminen, A. and Kaarniranta, K. (2009) Regulation of the aging process by autophagy. Trends Mol. Med. **15**, 217–224
37. Tavernarakis, N., Pasparaki, A., Tasdemir, E., Maiuri, M.C. and Kroemer, G. (2008) The effects of p53 on whole organism longevity are mediated by autophagy. Autophagy **4**, 870–873
38. Wong, E. and Cuervo, A.M. (2010) Autophagy gone awry in neurodegenerative diseases. Nat. Neurosci. **13**, 805–811
39. Harris, H. and Rubinsztein, D.C. (2012) Control of autophagy as a therapy for neurodegenerative disease. Nat. Rev. Neurol. **8**, 108–117
40. Zheng, S., Clabough, E.B., Sarkar, S., Futter, M., Rubinsztein, D.C. and Zeitlin, S.O. (2010) Deletion of the huntingtin polyglutamine stretch enhances neuronal autophagy and longevity in mice. PLoS Genet. **6**, e1000838
41. Lee, J., Giordano, S. and Zhang, J. (2012) Autophagy, mitochondria and oxidative stress: cross-talk and redox signalling. Biochem. J. **441**, 523–540
42. Lee, J.H., Yu, W.H., Kumar, A., Lee, S., Mohan, P.S., Peterhoff, C.M., Wolfe, D.M., Martinez-Vicente, M., Massey, A.C., Sovak, G. et al. (2010) Lysosomal proteolysis and autophagy require presenilin 1 and are disrupted by Alzheimer-related PS1 mutations. Cell **141**, 1146–1158
43. Tung, Y.T., Wang, B.J., Hu, M.K., Hsu, W.M., Lee, H., Huang, W.P. and Liao, Y.F. (2012) Autophagy: a double-edged sword in Alzheimer's disease. J. Biosci. **37**, 157–165

11

Role of autophagy in cancer prevention, development and therapy

G. Vignir Helgason*[1], Tessa L. Holyoake* and Kevin M. Ryan†[1]

*Paul O'Gorman Leukaemia Research Centre, Institute of Cancer Sciences, College of Medical, Veterinary & Life Sciences, University of Glasgow, Glasgow G12 0ZD, U.K., and †Tumour Cell Death Laboratory, Cancer Research UK Beatson Institute, Garscube Estate, Glasgow G61 1BD, U.K.

Abstract

Autophagy is a process that takes place in all mammalian cells and ensures homoeostasis and quality control. The term autophagy [self (auto)-eating (phagy)] was first introduced in 1963 by Christian de Duve, who discovered the involvement of lysosomes in the autophagy process. Since then, substantial progress has been made in understanding the molecular mechanism and signalling regulation of autophagy and several reviews have been published that comprehensively summarize these findings. The role of autophagy in cancer has received a lot of attention in the last few years and autophagy modulators are now being tested in several clinical trials. In the present chapter we aim to give a brief overview of recent findings regarding the mechanism and key regulators of autophagy and discuss the important physiological role of mammalian autophagy in health and disease. Particular focus is given to the role of autophagy in cancer prevention, development and in response to anticancer therapy. In this regard, we also give an updated list and discuss current clinical trials that aim to modulate autophagy, alone or in combination with radio-, chemo- or targeted therapy, for enhanced anticancer intervention.

Keywords:
anticancer therapy, autophagosome, chloroquine, chronic myeloid leukaemia, haemopoietic stem cell, hydroxychloroquine, tumorigenesis.

[1]Correspondence may be addressed to either of these authors (Vignir.Helgason@Glasgow.ac.uk or k.ryan@beatson.gla.ac.uk).

Introduction

Macroautophagy (hereafter referred to as autophagy) is an evolutionarily conserved catabolic process that involves degradation of cellular components in lysosomes. Recycling of these intracellular components can serve as an alternative source of energy during periods of metabolic stress or starvation (e.g. growth factor/nutrient deprivation) to maintain cellular homoeostasis and survival. During the autophagy process, double-membraned vesicles, termed autophagosomes, engulf long-lived proteins and organelles and transport these cargoes to lysosomes (Figure 1). Following fusion of the outer membrane of the autophagosome with the lysosomal membrane, the inner membrane, along with the cargo, is degraded by lysosomal hydrolases. Although autophagosomes in mammalian cells have been visualized by electron microscopy since as early as the 1950s, most of the molecular regulators of autophagy have only been characterized in the last two decades, following identification of the *ATG*

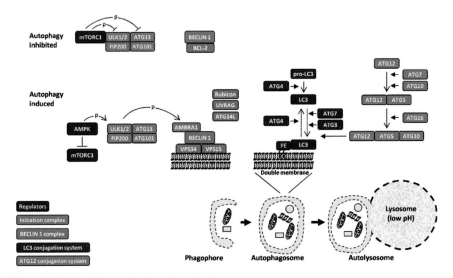

Figure 1. Molecular regulation of autophagy
Under normal nutrient-rich conditions, autophagy can be inhibited following phosphorylation of ULK1 and ATG13 by active mTORC1, or by Bcl-2-mediated inhibition of Beclin1. Induction of autophagy can occur following cellular stress or starvation that leads to mTORC1 dissociation from the ULK1 complex (that involves ULK1/2, ATG13, FIP200 and ATG101). AMPK induces autophagy by phosphorylating ULK1 on active sites. Active ULK1 phosphorylates AMBRA1, freeing Beclin1 to associate with membranes of the phagophore. Beclin1 interacts with several cofactors, such as AMBRA1, leading to activation of the lipid kinase VPS34, a class III PI3K that is critical for the expansion of phagophores. Beclin1 is also associated with UVRAG, ATG14L and Rubicon that are involved in the regulation of autophagy. Autophagosome maturation depends on two ubiquitin-like conjugation systems: (i) the LC3-conjugation system that starts when ATG4 cleaves the C-terminal of LC3 that is then activated by ATG7 (E1-like enzyme), transferred to ATG3 (E2-like enzyme) and eventually conjugated to PE. ATG4 also cleaves the amide bond between LC3 and PE to release the protein from membranes. (ii) The ATG12-conjugation system that starts with conjugation of ATG5 with ATG12 by ATG7 and ATG10 (E2-like enzyme). ATG12–ATG5 interacts then with ATG16 to form an E3-like complex that completes the LC3-conjugation reaction. During the autophagy process, double-membraned autophagosomes engulf proteins and organelles. Completion of the process involves fusion of the outer membrane of autophagosomes with lysosomes to form autolysosomes, where the inner membrane, along with the cargo, is degraded.

© The Authors Journal compilation © 2013 Biochemical Society

(autophagy-related) genes [1]. Most of these genes were initially identified in the budding yeast *Saccharomyces cerevisiae* and several of them have now been shown to have functional orthologues in mammalian cells. The evolutionarily conserved mechanism of autophagy has been thoroughly investigated and many detailed reviews have been published regarding the protein complexes that contribute to each step [2–4]. In the next section we briefly discuss the main regulators of autophagy and some of the signalling cascades, such as the RAS and PI3K (phosphoinositide 3-kinase)–Akt–mTORC1 [mammalian (also known as mechanistic) target of rapamycin complex 1] pathway, AMPK (AMP-activated protein kinase), p53 and Bcl-2 family proteins, which fine-tune its function and activity in a context-specific manner (also discussed in more detail in other chapters in this volume).

Molecular regulators of mammalian autophagy

Basal autophagy levels are usually low under normal conditions, but can be induced following cellular stress. The induction process starts with the regulated assembly of the initiation complex (sometimes referred to as the ULK1 complex) [5], a serine/threonine kinase complex containing ULK1 (uncoordinated-51-like kinase 1) and ULK2 (yeast Atg1), ATG13, FIP200 (focal adhesion kinase family-interacting protein 200 kDa; a mammalian functional homologue of Atg17) and ATG101 (Figure 1). These proteins associate with the phagophore (also known as the isolation membrane) upon autophagy induction.

A central inhibitor of autophagy in mammalian cells is the serine/threonine protein kinase complex mTORC1. mTORC1, which is activated under growing conditions (nutrient-rich), inhibits membrane targeting of the initiation complex by phosphorylating the ULK1 and ATG13 proteins of the complex [6]. Following mTORC1 inhibition, for example induced by starvation or rapamycin treatment, mTORC1 dissociates from the ULK1 complex and AMPK is free to phosphorylate ULK1 directly, allowing the complex to associate with membranes resulting in the initiation of autophagosome formation [7].

The essential autophagy protein Beclin1 (yeast Atg6) is also a key player in autophagosome biogenesis and interacts with several cofactors leading to activation of the lipid kinase VPS34, a class III PI3K, that is critical for the expansion of phagophores to double-membraned autophagosomes [8]. Under normal conditions, Beclin1 is, via its BH3 domain (Bcl-2 homology 3 domain), bound to and inhibited by Bcl-2 or the Bcl-2 homologue Bcl-xL [9]. BH3-only proteins and pharmacological BH3 mimetics can competitively disrupt the interaction between Beclin1 and Bcl-2/Bcl-xL to induce autophagy [10]. Beclin1 can also be found in complex with AMBRA1 (activating molecule in BECN1-regulated autophagy) and phosphorylation of AMBRA1 by ULK1 upon autophagy induction frees Beclin 1 for membrane association. Other Beclin1-binding partners, like UVRAG (UV radiation resistance-associated gene), ATG14L and Rubicon, have been shown to interact with Beclin1 and positively or negatively regulate autophagy [2].

The expansion of phagophores to autophagosomes and maturation of autophagosomes depend on two ubiquitin-like conjugation systems both of which are essential for autophagy. Each system is composed of two ubiquitin-like proteins, ATG8 (also known as microtubule-associated protein 1 light-chain 3, hereafter called LC3) and ATG12, and three enzymes (ATG3, ATG7 and ATG10) that are required for conjugation reactions [11]. The conjugation system of LC3 starts when ATG4 cleaves the C-terminus of LC3. LC3 is then activated by ATG7 (E1-like enzyme), transferred to ATG3 (E2-like enzyme) and eventually conjugated to PE (phosphatidylethanolamine). ATG4 also cleaves the amide bond between LC3 and PE to release the protein from membranes. The conjugation system of ATG12 starts with conjugation of ATG5 with ATG12 by ATG7, in a similar manner to the conjugation reaction of LC3, except now ATG10 functions as the E2 enzyme instead of ATG3. The ATG12–ATG5 conjugate interacts with ATG16 to form a complex that exerts an E3 enzyme-like function on the LC3 conjugation reaction [12]. Lipidated LC3 is most commonly used to monitor autophagy by various assays although other techniques are also used (researchers in the field have recently updated the comprehensive guidelines for monitoring autophagy [13]).

The completion process of autophagy includes the fusion of the outer membrane of autophagosomes with lysosomes, which contain pH-sensitive degradative enzymes, to form autolysosomes, although the mechanistic details for this step are still not completely clear. Following fusion, the inner single membrane of the autophagosome and its cargo are lysed by lysosomal hydrolases, especially cathepsins, and degraded; this process has been demonstrated to require the lysosomal protein LAMP-2, the small GTPase RAB7 and UVRAG [14–17].

Signalling regulation of mammalian autophagy

The multistep autophagic pathway is tightly controlled by several signalling mechanisms such as growth factor signalling, energy sensing, ER (endoplasmic reticulum) stress, hypoxia, oxidative stress and pathogen infection [2]. Below, we briefly discuss how some of these signalling pathways, such as the RAS–RAF–MEK1/2 [MAPK (mitogen-activated protein kinase)/ERK (extracellular-signal-regulated kinase) kinase 1/2]–ERK1/2, PI3K–Akt–mTORC1 and p53, which are frequently deregulated in cancer and are important for tumorigenesis, can modulate autophagy (Figure 2).

Growth factor signalling: RAS and the PI3K–Akt–mTORC1 pathway

A growing body of evidence suggests that increased mTOR activity is important for cancer pathogenesis and may provide cancer cells with the machinery required to sustain high levels of cell growth [18]. mTOR exists in two conserved protein complexes, mTORC1 and mTORC2. mTORC1 has a primary function in regulating autophagy, acts as a nutrient sensor and has been described as the master regulator of autophagy [19]. Upstream of mTORC1 are the class I PI3K and Akt kinases that can lead to inhibition of autophagy when active. PI3K, a factor commonly mutated in human cancers, catalyses the production of PIP_3 (phosphatidylinositol 3,4,5-trisphosphate) at the plasma membrane, which in turn increases membrane recruitment of Akt. Active Akt activates mTORC1 by inhibiting a downstream protein complex, the TSC (tuberous sclerosis complex)1–TSC2. The TSC1–TSC2 complex functions as a GTPase-activating protein for Rheb (Ras homologue enriched in brain), a small GTP-binding protein that binds to and activates mTORC1. The TSC1–TSC2 complex also senses inputs from other kinases such as ERK1/2. It has been shown that phosphorylation of TSC2 by Akt or ERK1/2

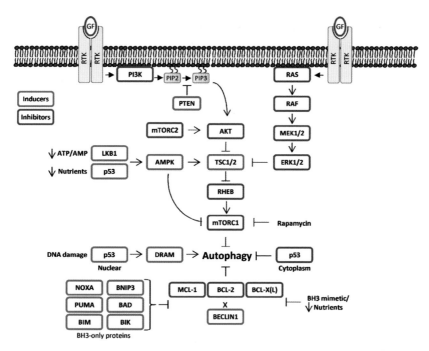

Figure 2. Signalling regulation of autophagy
Autophagy is tightly controlled by several signalling pathways. GF (growth factor)-mediated stimulation of RTKs (receptor tyrosine kinases) can inhibit autophagy through activation of the PI3K–Akt–mTORC1 or the RAS–RAF–MEK–ERK pathways. Active PI3K phosphorylates membrane-bound PIP_2 to form PIP_3, a process that can be inhibited by the PTEN (phosphatase and tensin homologue deleted on chromosome 10) tumour suppressor protein. PIP_3 increases membrane recruitment of Akt that activates mTORC1 by inhibiting the TSC1–TSC2 complex (a GTPase-activating protein for Rheb). The TSC1–TSC2 complex can also be inhibited by ERK1/2 resulting in mTORC1 activation. On the other hand, decreased ATP/AMP ratio or glucose starvation can lead to AMPK activation via LKB1 kinase or p53, leading to activation of the TSC1–TSC2 complex and through Rheb, inhibition of mTORC1, resulting in autophagy induction. mTORC1 can also be directly inhibited by rapamycin treatment, leading to autophagy induction. Nuclear p53 has been shown to induce DRAM-1-dependent autophagy, whereas cytoplasmic p53 has been shown to inhibit autophagy, indicating a dual and complex role of p53 in autophagy control. The Bcl-2 family proteins have also been shown to play a dual role in autophagy regulation; the anti-apoptotic proteins Bcl-2, Bcl-xL and MCL-1, can inhibit autophagy by binding to and inhibiting Beclin1, whereas the pro-apoptotic BH3-only proteins (Noxa, PUMA, BIM, BNIP3, BAD and BIK) can induce autophagy by interfering with the anti-apoptotic protein/Beclin1 association. BH3 mimetics or reduction in nutrients has also been shown to induce autophagy through inhibition of the Bcl-2/Beclin1 association.

leads to the disruption of the complex, resulting in mTORC1 activation [20]. The role of RAS in regulating autophagy is therefore complex as it has been documented to have opposing roles; it can inhibit autophagy by activating ERK1/2 or the PI3K–Akt–mTORC1 pathway [21], or it may activate the RAF–MEK1/2–ERK1/2 pathway leading to phosphorylation of the GIAP (Gα-interacting protein) (protein involved in G-protein signalling regulation) and autophagy induction [22]. This is in line with data suggesting that human cancer cell lines bearing activating mutations in RAS can have high levels of autophagy that is required to maintain oxidative metabolism and tumorigenesis [23].

Energy sensing: positive regulation by AMPK

AMPK is an energy stress sensor that is activated by LKB1 kinase during nutrient and energy depletion (decreased ATP/AMP ratio). In contrast with active Akt and ERK1/2, active AMPK leads to phosphorylation and activation of the TSC1–TSC2 complex, and through Rheb, to inhibition of mTORC1. Thus AMPK is a positive regulator of autophagy. AMPK can also inhibit mTORC1 activity by directly phosphorylating raptor (regulatory associated protein of mTOR), a subunit of mTORC1, and this phosphorylation may be crucial for the inhibition of mTORC1 [24].

Stress response: the dual role of p53

The p53 tumour suppressor is frequently mutated in the majority of human cancers [25], and has been shown to have a dual role in autophagy control. p53 has been shown to induce autophagy in a manner dependent on one of its target genes, DRAM-1 (damage-regulated autophagy modulator 1) [26]. DRAM-1 was also found to be involved in p53-induced death, highlighting the relationship between cell death, p53 and autophagy. p53 has also been shown to induce autophagy through activation of AMPK/inhibition of mTORC1 following glucose starvation [27]. On the other hand, both cytoplasmic wild-type and mutant p53 proteins have been shown to inhibit autophagy [28], in contrast with the pro-autophagic activity of nuclear p53.

The link to cell death: Bcl-2 protein family

Bcl-2 is the prototype of a family of proteins containing at least one BH domain. The Bcl-2 family proteins are subdivided into three groups on the basis of their pro- or anti-apoptotic action and the BH domains they possess: first, anti-apoptotic multidomain Bcl-2-like proteins [such as Bcl-2, Bcl-xL and MCL-1 (myeloid cell leukaemia sequence 1)], which contain four BH domains (BH1–BH4). Secondly, pro-apoptotic multidomain BAX-like proteins (such as BAX and BAK), which also contain four BH domains, and thirdly, the pro-apoptotic BH3-only protein family [such as BID (Bcl-2 homology 3 interacting-domain death agonist), BIM (Bcl-2-interacting mediator of cell death), BAD (Bcl-2/Bcl-xL-antagonist, causing cell death), Noxa and PUMA (p53 up-regulated modulator of apoptosis)] [29]. BH3-only proteins affect cell death by interacting with the anti-apoptotic Bcl-2-like proteins, which possess a hydrophobic cleft (the BH3-binding groove), to inhibit their function and/or by interacting directly with pro-apoptotic multidomain proteins (BAX and BAK) to stimulate their activity. Following activation, BAX and BAK associate with the mitochondrial membrane where they cause outer membrane permeabilization. This induces the release of cytochrome *c* and other pro-apoptotic factors from the mitochondria, leading to activation of caspases and apoptosis. In addition to cell death regulation, the Bcl-2 protein family also plays a dual role in autophagy regulation. The anti-apoptotic Bcl-2, Bcl-xL and MCL-1 can inhibit autophagy, whereas the pro-apoptotic BH3-only proteins BNIP3, BAD, BIK (Bcl-2-interacting killer), Noxa, PUMA and BIM can induce autophagy [10]. As mentioned earlier, the binding of Bcl-2 to Beclin1 disrupts the association of Beclin1 with VPS34, decreasing VPS34 activity and thereby inhibiting autophagy. The Beclin1–Bcl-2 interaction has been shown to be reduced following starvation or treatment with pharmacological BH3 mimetics, freeing Beclin1 to engage in autophagy activation [30].

The paradoxical role of autophagy in cancer

The first strong link between autophagy and cancer was published in 1999 when it was shown that Beclin1 can inhibit tumorigenesis and that its levels are diminished in human breast carcinoma, suggesting that decreased expression of autophagy proteins may contribute to the development or progression of human cancer [31]. Subsequent studies revealed that *Beclin1* is a haploinsufficient tumour suppressor and essential for early embryonic development [32,33]. In line with this notion are studies demonstrating that frameshift mutations of ATG2B, ATG5, ATG9B and ATG12 can be found in human cancers [34] and that deletion of other key autophagy regulators in mice can push cells towards malignant transformation [35,36]. In this regard, recent studies on the role of basal autophagy on the survival and function of HSCs (haemopoietic stem cells) have revealed that autophagy plays a critical role for HSC maintenance [37] and may protect against leukaemia development. The first indication came in 2010 when it was shown that FIP200 is required for the maintenance and function of fetal HSCs and deletion of *FIP200* resulted in increased mitochondrial mass and high levels of ROS (reactive oxygen species) followed by severe anaemia and perinatal lethality [38]. In another study, conditional deletion of *Atg7* in the haemopoietic system resulted in an accumulation of mitochondria, ROS and DNA damage, followed by loss of HSC function [35]. Moreover, the production of both lymphoid and myeloid progenitors was impaired in the absence of *Atg7* and the mice developed a myeloproliferative disorder and died within weeks, indicating that autophagy may protect against leukaemogenesis. Taken together, it is now evident that basal autophagy within normal cells, particularly stem cells, is pivotal since it functions as a 'guardian' by promoting adaptation under changing conditions and/or stress, maintains protein/organelle quality control and metabolism, prevents accumulation of p62 [39], regulates removal of damaged mitochondria (that would otherwise produce ROS and damage the DNA), and therefore promotes genetic stability [40].

On the other hand, increasing evidence suggests that autophagy also acts as a survival mechanism within cancer cells. Leukaemia stem cells living in the bone marrow and other cancer cells, particularly cells in the core of solid tumours, have to overcome adverse conditions such as hypoxia and limited access to the vascular niche and nutrients and may therefore rely more heavily on autophagy than normal cells. Moreover, autophagy appears to be particularly important for cancer cell survival following stress or anticancer therapy. In line with this, we have shown that IM (imatinib) mediated Bcr-Abl inhibition, which has been reported to promote apoptosis by inducing the expression or activity of the BH3-only protein BIM and BAD [41], induces protective autophagy in CML (chronic myeloid leukaemia) cells [42]. Of importance, specific autophagy inhibition, either with *ATG7* or *ATG5* knockdown, or pharmacological inhibition using CQ (chloroquine), results in enhanced death induced by IM, in cell lines and primary CML stem cells. Furthermore, inhibition of autophagy has been shown to lead to enhanced apoptosis following irradiation [43], p53 activation [44] and following treatment with other apoptosis activators such as alkylating agents [44] and BH3 mimetics [45,46]. This has provided a rationale for initiation of clinical trials where the use of autophagy inhibitors in combination with chemotherapy or targeted therapy is being tested (Tables 1 and 2).

Table 1. Drugs used in combination with HCQ-mediated autophagy inhibition in currently active clinical trials

EGFR, epidermal growth factor receptor; HDAC, histone deacetylase; RTK, receptor tyrosine kinase; TK, tyrosine kinase.

Intervention	Drug	Target
Chemotherapy	5-Fluorouracil	Thymidylate synthase
	Capecitabine	DNA synthesis
	Carboplatin	DNA
	Gemcitabine	DNA replication
	Leucovorin	Chemoprotectant
	Oxaliplatin	DNA synthesis
	Paclitaxel	Microtubules
	Temozolomide	DNA (alkylating agent)
	Cyclophosphamide	DNA (alkylating agent)
Targeted therapy	Bevacizumab	Angiogenesis
	Bortezomib	Proteasome
	Erlotinib	TK
	Gefitinib	EGFR
	Imatinib	TK
	MK2206	Akt
	RAD001	mTOR
	Sirolimus/rapamycin	mTOR
	Sorafenib	TK
	Sunitinib malate	RTK
	Temsirolimus	mTOR
	Vorinostat	HDAC

Table 2. Selective active clinical trials where autophagy inhibition is being tested

BWH, Brigham and Women's Hospital; CRUK, Cancer Research UK; DFCI, Dana Farber Cancer Institute; DM, Devalingam Mahalingam; LM, Lynn McMahon; MAASTRO, Maastricht Radiation Oncology; MGH, Massachusetts General Hospital; MP, Millennium Pharmaceuticals; MRC, Medical Research Council; MUMC, Maastricht University Medical Center; NCI, National Cancer Institute; NIH, National Institutes of Health; NYU, New York University; RTK, receptor tyrosine kinase; SKCCC, Sidney Kimmel Comprehensive Cancer Center; TKI, tyrosine kinase inhibitor; UCL, University College, London; UMDNJ, University of Medicine and Dentistry New Jersey; UPenn, University of Pennsylvania; VCU, Virginia Commonwealth University.

Condition	Intervention	Phase	Sponsor, collaborator	Identifier
Adult solid tumour	HCQ + radiotherapy	I	VCU, NCI	NCT01417403
Advanced cancers	HCQ + temsirolimus	I	UPenn, NCI	NCT00909831
	HCQ + sirolimus or vorinostat	I	M.D. Anderson Cancer Center	NCT01266057
Advanced solid tumours	HCQ + temozolomide	I	UPenn, NCI	NCT00714181
	HCQ + MK2206	I	NCI	NCT01480154
	HCQ + vorinostat	I	DM, Merck	NCT01023737
	HCQ + sunitinib malate	I	NCI	NCT00813423
Breast cancer	HCQ	II	Radboud University	NCT01292408
CML	HCQ + IM mesylate	I	LM, MRC, CRUK Trials unit Glasgow	NCT01227135

(Continued)

Table 2. Selective active clinical trials where autophagy inhibition is being tested *(Continued)*

Condition	Intervention	Phase	Sponsor, collaborator	Identifier
Colorectal cancer	HCQ + bevacizumab, capecitabine, oxaliplatin	II	UMDNJ, NCI	NCT01006369
	HCQ + oxaliplatin, leucovorin, 5-fluorouracil, bevacizumab	I/II	UPenn	NCT01206530
Glioblastoma multiforme	HCQ + radiotherapy	I	UCL, CRUK	NCT01602588
	HCQ + temozolomi	I/II	SKCCC, NCI	NCT00486603
Metastatic pancreatic cancer	HCQ	II	DFCI, BWH, MGH	NCT01273805
Multiple myeloma	HCQ + bortezomib	I/II	UPenn, NCI	NCT00568880
	HCQ + rapamycin	I	OHSU Knight Cancer Institute	NCT01689987
	CQ + cyclophosphamide, bortezomib	II	NYU School of Medicine, MP	NCT01438177
Non-small-cell lung cancer	HCQ + gefitinib	I/II	NUH, MGH, AstraZeneca	NCT00809237
	HCQ + paclitaxel, carboplatin, bevacizumab	I/II	UMDNJ, NCI	NCT00933803

Pancreatic cancer	HCQ + erlotinib	II	MGH, Stanford University, Yale University, University of Maryland, Genentech	NCT00977470
	HCQ + paclitaxel, carboplatin, bevacizumab	II	UMDNJ, NCI	NCT01649947
	HCQ + capecitabine, radiotherapy	II	MGH	NCT01494155
	HCQ + gemcitabine	I/II	UPenn	NCT01506973
	HCQ + gemcitabine	I/II	University of Pittsburgh, NIH	NCT01128296
Prostate cancer	HCQ	II	UMDNJ, NCI	NCT00726596
Renal cell carcinoma	HCQ	I	University of Pittsburgh, NIH	NCT01144169
	HCQ + RAD001	I/II	UPenn	NCT01510119
Small-cell lung cancer	CQ	I	MAASTRO, MUMC, NCI	NCT00969306
Solid tumours	HCQ + sorafenib	I	Tyler Curiel	NCT01634893

Autophagy therapeutics

Given the paradoxical role of autophagy in cancer, it is not surprising that autophagy modulation has been warned to be a double-edged sword (that is, it could have both favourable and unfavourable consequences) [47]. Establishing how the functional status of autophagy may influence treatment response is therefore critical. Below, we discuss some examples of how autophagy induction could be beneficial and when autophagy inhibition could lead to improved therapy.

Autophagy induction in cancer prevention and improved chemotherapy

If autophagy serves to preserve cellular integrity, could boosting autophagy above basal levels cause an even greater protection against tumour development? CR (caloric restriction), i.e. reduced food intake without malnutrition, affects molecular pathways that are also known to be altered in cancer. This can lead to inhibition of mTOR through activation of AMPK or SIRT1, or inhibition of growth factor signalling, and is probably the most physiological inducer of autophagy. Indeed, it has been shown that CR can be effective in cancer prevention in animal models [48–50] and reduce cancer incidence by 50% in monkeys [51]. Epidemiological studies suggest that CR is also beneficial to human health [52]. Whether CR is effective in reducing the risk for developing cancer, such as breast cancer, is now being studied in the clinic (http://clinicaltrials.gov/; search for 'CR and cancer'). However, whether CR will only prevent cancer from developing in diseases for which being overweight is an accepted risk factor, and if the effect relies partly on autophagy, is still unclear.

Furthermore, is it possible that autophagy induced by lowering caloric intake can prevent tumour progression or even enhance anticancer treatment, either by sensitizing cancer cells to death or by protecting normal cells from highly toxic drug treatment? Interestingly, CR has been shown to decrease tumour progression in p53-deficient mice [53] and reduce the growth of mammary tumours [54]. In addition, CR has been shown to have a similar effect to rapamycin treatment by leading to decreased mTORC1 activity and reduced tumour volume in a murine model of pancreatic cancer [55], suggesting that autophagy may play a role in slowing tumour progression.

Fasting, another way to induce autophagy, has been shown recently to sensitize cancer cells to radiotherapy or chemotherapy and led to extended survival in *in vivo* GBM (glioblastoma multiforme) models, indicating that fasting could enhance the efficacy of existing cancer treatments in patients [56]. Fasting has also been shown to protect normal, but not cancer, cells against chemotherapy [57], indicating that autophagy modulation may have the potential to maximize the differential toxicity of normal and cancer cells. However, the mechanism for this is still not clear, and whether this effect relies on fasting-induced autophagy needs to be further examined. In fact, initiation of several clinical studies (the ClinicalTrials database; search for 'fasting and cancer') suggests that answers regarding whether autophagy-inducing measures may be effective at reducing tumour growth or the toxicity of chemotherapy in humans are on the horizon.

Autophagy inhibition in combination with anticancer therapy

In the clinic, autophagy can be inhibited using the antimalarial drug HCQ (hydroxychloroquine), which has also been approved for treatment of a variety of disorders. HCQ is not a specific autophagy inhibitor, but a lysosomotropic weak base, that raises the pH within lysosomes, impairs lysosomal function and therefore autophagic protein degradation. Despite the controversy regarding the role of autophagy in cancer, since HCQ has been used extensively in the treatment of malaria and rheumatoid arthritis and is quite well tolerated, close to 30 clinical trials are ongoing in cancer patients using HCQ alone or in combination with cytotoxic agents (Table 2). Our promising *in vitro* data on the effect of IM/CQ combination on survival of CML stem cells [42] have led to a Phase II clinical trial (CHOICES; CHlOroquine and IM Combination to Eliminate Stem cells), the first clinical trial testing autophagy inhibition in CML [58]. The CHOICES trial aims to test the combination of IM with HCQ in IM-sensitive CML patients who continue to show evidence for residual disease caused by the persistence of CML stem cells (Figure 3). Since CML has long been described as a paradigm for targeted therapy with the potential to provide a cure for cancer patients, it is hoped that the CHOICES trial will not only answer the question as to whether autophagy can be effectively inhibited in

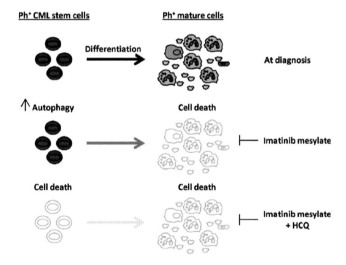

Figure 3. IM combined with HCQ for elimination of persistent CML stem cells
CML is a stem cell disorder. The hallmark is the Philadelphia (Ph) chromosome, which forms as a result of translocation between chromosome 9 and 22 in an HSC, leading to expression of the Bcr-Abl fusion protein. This leads to expansion and differentiation of myeloid cells causing the disease. IM, a TKI (tyrosine kinase inhibitor) that inhibits Bcr-Abl activity and used as first-line treatment for CML, is effective in killing differentiated Ph+ CML cells; however, Ph+ CML stem cells are insensitive to IM, leading to disease persistence in the majority of patients. IM has been shown to induce autophagy in CML stem cells and CQ-mediated autophagy inhibition resulted in near complete elimination of CML stem cells *in vitro*. That has led to CHOICES, a Phase II clinical trial that aims to test a IM/HCQ combination in CML patients who continue to show evidence for residual disease caused by the persistence of CML stem cells following IM treatment.

patients, but also if cancer stem cells/cancer-initiating cells may prove vulnerable to autophagy inhibition, and therefore provide a platform for other stem-cell-driven cancer studies. In line with our data, recent work by Galavotti et al. [59] on the role of autophagy in GBM stem cells showed that DRAM-1 regulates invasive properties of GBM stem cells and its expression is associated with shorter overall survival in GBM patients. Two trials are also currently ongoing to test whether treatment with HCQ in combination with radiation is beneficial in patients with GBM.

As mentioned earlier, since complete inhibition of autophagy using mouse models has significant detrimental effects on survival [32,35,38] and given the tumour suppressive function of autophagy and its role in responses to chemotherapy, is there a potential risk associated with autophagy inhibition in humans? This has been suggested in a previous study by Michaud et al. [60], who showed that autophagy inhibition may limit chemotherapy responses by preventing autophagy-dependent anticancer immune responses, raising the concern that acute autophagy inhibition may actually limit chemotherapy responses in certain cancers for which an immune reaction plays an important role in disease response. However, if treatment regimens involve short-lived and/or impairment as opposed to complete inhibition of autophagy, this may limit toxicity and the propensity to develop secondary malignancies while still achieving therapeutic gain.

Future directions

Despite the numerous clinical trials, it is still early days for autophagy inhibition in the clinical setting. Most patients are treated with 200–800 mg of HCQ daily, which in most cases is well tolerated; however, although 1–10 µM concentration HCQ inhibits autophagy *in vitro*, it is still not clear whether autophagy is blocked in a sustained fashion in all patients receiving up to 800, or even 1200 mg daily. This therefore questions whether any therapeutic effects of HCQ or CQ are even partly related to the inhibition of autophagy. In this regard, recent *in vitro* studies showed that the ability of CQ to enhance chemotherapeutic responses was the same whether the cells were autophagy competent or deficient [61]. It may therefore be essential to develop more potent/specific autophagy inhibitors for use in future clinical trials. Recently, Lys05, a CQ derivative that is ten times more potent than HCQ [62], spautin-1, a small molecule inhibitor of VPS34 [63] and clarithromycin (a form of macrolide antibiotic) [64] have shown promising results in pre-clinical models or in patients [65].

Conclusions

The role of autophagy in cancer has been widely studied in recent years and current knowledge suggests that autophagy can both (i) protect normal cells from accumulation of damaged DNA/proteins and therefore prevent tumour formation, and (ii) help cancer cells to adapt to a hostile environment and protect them from anticancer therapy. Despite this dual role of autophagy, as many studies support the cytoprotective role of autophagy in cancer cells, inhibition of the process is being tested in numerous clinical trials using HCQ, a non-specific autophagy inhibitor and more specific/potent inhibitors are in pre-clinical development. With development of novel autophagy-modulating agents, in addition to greater understanding of the tissue/cancer-specific role of autophagy, it is hoped that autophagy may provide a therapeutic potential for cancer and other human diseases in the future.

Summary
- Autophagy is an evolutionarily conserved recycling process that takes place in every mammalian cell.
- Many *ATG* (autophagy-related) genes have been characterized that regulate the process at the molecular level; from initiation/formation of autophagosome (ULK1 complex), to expansion (VPS34–Beclin1 complex), maturation (LC3 and ATG12 conjugation systems) and finally completion/degradation (fusion with lysosomes).
- Under normal conditions autophagy levels are usually low, but can be induced following cellular stress or starvation, often involving inhibition of the PI3K–Akt–mTORC1 pathway.
- Autophagy has dual roles in cancer; it has a tumour suppressive function by preventing DNA damage and genomic instability, yet it promotes tumour development by promoting cancer cell survival under diverse conditions.
- CR has been shown to both prevent tumour progression and enhance anticancer treatment, where autophagy has been suggested to play a role.
- Autophagy is induced in cancer cells following radiation or treatment with various apoptosis activators.
- Autophagy inhibition is being tested in the clinic, using HCQ, a non-specific autophagy inhibitor, alone but mostly in combination with chemotherapy and radiotherapy.

We thank Maria Karvela for reading over the chapter. We apologize that many important references have been omitted because of limitations on the number of citations. G.V.H. is funded by the Kay Kendall Leukaemia Fund (KKL404 and KKL698) and Lord Kelvin Adam Smith Leadership Fellowship scheme, University of Glasgow. Work in T.L.H.'s laboratory is supported by the MRC (Medical Research Council) (G0900882, CHOICES, ISCRTN No. 61568166), University of Glasgow, SULSA, SNBTS and Cancer Research UK programme funding. Work in the Tumour Cell Death Laboratory is supported by grants from Cancer Research UK and the Association for International Cancer Research to K.M.R.

References
1. Klionsky, D.J. (2007) Autophagy: from phenomenology to molecular understanding in less than a decade. Nat. Rev. Mol. Cell Biol. **8**, 931–937
2. Yang, Z. and Klionsky, D.J. (2010) Mammalian autophagy: core molecular machinery and signaling regulation. Curr. Opin. Cell Biol. **22**, 124–131

3. Xie, Z. and Klionsky, D.J. (2007) Autophagosome formation: core machinery and adaptations. Nat. Cell Biol. **9**, 1102–1109
4. Mizushima, N. (2007) Autophagy: process and function. Genes Dev. **21**, 2861–2873
5. Mizushima, N. (2010) The role of the Atg1/ULK1 complex in autophagy regulation. Curr. Opin. Cell Biol. **22**, 132–139
6. Hosokawa, N., Hara, T., Kaizuka, T., Kishi, C., Takamura, A., Miura, Y., Iemura, S., Natsume, T., Takehana, K., Yamada, N. et al. (2009) Nutrient-dependent mTORC1 association with the ULK1–Atg13–FIP200 complex required for autophagy. Mol. Biol. Cell **20**, 1981–1991
7. Jung, C.H., Jun, C.B., Ro, S.H., Kim, Y.M., Otto, N.M., Cao, J., Kundu, M., and Kim, D.H. (2009) ULK–Atg13–FIP200 complexes mediate mTOR signaling to the autophagy machinery. Mol. Biol. Cell **20**, 1992–2003
8. Simonsen, A. and Tooze, S.A. (2009) Coordination of membrane events during autophagy by multiple class III PI3-kinase complexes. J. Cell Biol. **186**, 773–782
9. Maiuri, M.C., Le Toumelin, G., Criollo, A., Rain, J.C., Gautier, F., Juin, P., Tasdemir, E., Pierron, G., Troulinaki, K., Tavernarakis, N. et al. (2007) Functional and physical interaction between Bcl-X(L) and a BH3-like domain in Beclin-1. EMBO J. **26**, 2527–2539
10. Levine, B., Sinha, S. and Kroemer, G. (2008) Bcl-2 family members: dual regulators of apoptosis and autophagy. Autophagy **4**, 600–606
11. Ohsumi, Y. (2001) Molecular dissection of autophagy: two ubiquitin-like systems. Nat. Rev. Mol. Cell Biol. **2**, 211–216
12. Geng, J. and Klionsky, D.J. (2008) The Atg8 and Atg12 ubiquitin-like conjugation systems in macroautophagy. 'Protein modifications: beyond the usual suspects' review series. EMBO Rep. **9**, 859–864
13. Klionsky, D.J., Abdalla, F.C., Abeliovich, H., Abraham, R.T., Acevedo-Arozena, A., Adeli, K., Agholme, L., Agnello, M., Agostinis, P., Aguirre-Ghiso, J.A. et al. (2012) Guidelines for the use and interpretation of assays for monitoring autophagy. Autophagy **8**, 445–544
14. Eskelinen, E.L., Tanaka, Y. and Saftig, P. (2003) At the acidic edge: emerging functions for lysosomal membrane proteins. Trends Cell Biol. **13**, 137–145
15. Gutierrez, M.G., Munafo, D.B., Beron, W. and Colombo, M.I. (2004) Rab7 is required for the normal progression of the autophagic pathway in mammalian cells. J. Cell Sci. **117**, 2687–2697
16. Jager, S., Bucci, C., Tanida, I., Ueno, T., Kominami, E., Saftig, P. and Eskelinen, E.L. (2004) Role for Rab7 in maturation of late autophagic vacuoles. J. Cell Sci. **117**, 4837–4848
17. Liang, C., Lee, J.S., Inn, K.S., Gack, M.U., Li, Q., Roberts, E.A., Vergne, I., Deretic, V., Feng, P., Akazawa, C. and Jung, J.U. (2008) Beclin1-binding UVRAG targets the class C Vps complex to coordinate autophagosome maturation and endocytic trafficking. Nat. Cell Biol. **10**, 776–787
18. Laplante, M. and Sabatini, D.M. (2012) mTOR signaling in growth control and disease. Cell **149**, 274–293
19. Jung, C.H., Ro, S.H., Cao, J., Otto, N.M. and Kim, D.H. (2010) mTOR regulation of autophagy. FEBS Lett. **584**, 1287–1295
20. Ma, L., Chen, Z., Erdjument-Bromage, H., Tempst, P. and Pandolfi, P.P. (2005) Phosphorylation and functional inactivation of TSC2 by Erk implications for tuberous sclerosis and cancer pathogenesis. Cell **121**, 179–193
21. Furuta, S., Hidaka, E., Ogata, A., Yokota, S. and Kamata, T. (2004) Ras is involved in the negative control of autophagy through the class I PI3-kinase. Oncogene **23**, 3898–3904
22. Pattingre, S., Bauvy, C. and Codogno, P. (2003) Amino acids interfere with the ERK1/2-dependent control of macroautophagy by controlling the activation of Raf-1 in human colon cancer HT-29 cells. J. Biol. Chem. **278**, 16667–16674
23. Guo, J.Y., Chen, H.Y., Mathew, R., Fan, J., Strohecker, A.M., Karsli-Uzunbas, G., Kamphorst, J.J., Chen, G., Lemons, J.M., Karantza, V. et al. (2011) Activated Ras requires autophagy to maintain oxidative metabolism and tumorigenesis. Genes Dev. **25**, 460–470

24. Gwinn, D.M., Shackelford, D.B., Egan, D.F., Mihaylova, M.M., Mery, A., Vasquez, D.S., Turk, B.E. and Shaw, R.J. (2008) AMPK phosphorylation of raptor mediates a metabolic checkpoint. Mol. Cell **30**, 214–226
25. Muller, P.A. and Vousden, K.H. (2013) p53 mutations in cancer. Nat. Cell Biol. **15**, 2–8
26. Crighton, D., Wilkinson, S., O'Prey, J., Syed, N., Smith, P., Harrison, P.R., Gasco, M., Garrone, O., Crook, T. and Ryan, K.M. (2006) DRAM, a p53-induced modulator of autophagy, is critical for apoptosis. Cell **126**, 121–134
27. Feng, Z., Zhang, H., Levine, A.J. and Jin, S. (2005) The coordinate regulation of the p53 and mTOR pathways in cells. Proc. Natl. Acad. Sci. U.S.A. **102**, 8204–8209
28. Tasdemir, E., Maiuri, M.C., Galluzzi, L., Vitale, I., Djavaheri-Mergny, M., D'Amelio, M., Criollo, A., Morselli, E., Zhu, C., Harper, F. et al. (2008) Regulation of autophagy by cytoplasmic p53. Nat. Cell Biol. **10**, 676–687
29. Martinou, J.C. and Youle, R.J. (2011) Mitochondria in apoptosis: Bcl-2 family members and mitochondrial dynamics. Dev. Cell **21**, 92–101
30. Pattingre, S., Tassa, A., Qu, X., Garuti, R., Liang, X.H., Mizushima, N., Packer, M., Schneider, M.D. and Levine, B. (2005) Bcl-2 antiapoptotic proteins inhibit Beclin 1-dependent autophagy. Cell **122**, 927–939
31. Liang, X.H., Jackson, S., Seaman, M., Brown, K., Kempkes, B., Hibshoosh, H. and Levine, B. (1999) Induction of autophagy and inhibition of tumorigenesis by beclin 1. Nature **402**, 672–676
32. Yue, Z., Jin, S., Yang, C., Levine, A.J. and Heintz, N. (2003) Beclin 1, an autophagy gene essential for early embryonic development, is a haploinsufficient tumor suppressor. Proc. Natl. Acad. Sci. U.S.A. **100**, 15077–15082
33. Qu, X., Yu, J., Bhagat, G., Furuya, N., Hibshoosh, H., Troxel, A., Rosen, J., Eskelinen, E.L., Mizushima, N., Ohsumi, Y. et al. (2003) Promotion of tumorigenesis by heterozygous disruption of the beclin 1 autophagy gene. J. Clin. Invest. **112**, 1809–1820
34. Kang, M.R., Kim, M.S., Oh, J.E., Kim, Y.R., Song, S.Y., Kim, S.S., Ahn, C.H., Yoo, N.J. and Lee, S.H. (2009) Frameshift mutations of autophagy-related genes ATG2B, ATG5, ATG9B and ATG12 in gastric and colorectal cancers with microsatellite instability. J. Pathol. **217**, 702–706
35. Mortensen, M., Soilleux, E.J., Djordjevic, G., Tripp, R., Lutteropp, M., Sadighi-Akha, E., Stranks, A.J., Glanville, J., Knight, S., Jacobsen, S.E. et al. (2011) The autophagy protein Atg7 is essential for hematopoietic stem cell maintenance. J. Exp. Med. **208**, 455–467
36. Takamura, A., Komatsu, M., Hara, T., Sakamoto, A., Kishi, C., Waguri, S., Eishi, Y., Hino, O., Tanaka, K. and Mizushima, N. (2011) Autophagy-deficient mice develop multiple liver tumors. Genes Dev. **25**, 795–800
37. Warr, M.R., Binnewies, M., Flach, J., Reynaud, D., Garg, T., Malhotra, R., Debnath, J. and Passegue, E. (2013) FOXO3A directs a protective autophagy program in haematopoietic stem cells. Nature **494**, 323–327
38. Liu, F., Lee, J.Y., Wei, H., Tanabe, O., Engel, J.D., Morrison, S.J. and Guan, J.L. (2010) FIP200 is required for the cell-autonomous maintenance of fetal hematopoietic stem cells. Blood **116**, 4806–4814
39. Mathew, R., Karp, C.M., Beaudoin, B., Vuong, N., Chen, G., Chen, H.Y., Bray, K., Reddy, A., Bhanot, G., Gelinas, C. et al. (2009) Autophagy suppresses tumorigenesis through elimination of p62. Cell **137**, 1062–1075
40. Mathew, R., Kongara, S., Beaudoin, B., Karp, C.M., Bray, K., Degenhardt, K., Chen, G., Jin, S. and White, E. (2007) Autophagy suppresses tumor progression by limiting chromosomal instability. Genes Dev. **21**, 1367–1381
41. Kuroda, J., Puthalakath, H., Cragg, M.S., Kelly, P.N., Bouillet, P., Huang, D.C., Kimura, S., Ottmann, O.G., Druker, B.J., Villunger, A. et al. (2006) Bim and Bad mediate imatinib-induced killing of Bcr/Abl+ leukemic cells, and resistance due to their loss is overcome by a BH3 mimetic. Proc. Natl. Acad. Sci. U.S.A. **103**, 14907–14912

42. Bellodi, C., Lidonnici, M.R., Hamilton, A., Helgason, G.V., Soliera, A.R., Ronchetti, M., Galavotti, S., Young, K.W., Selmi, T., Yacobi, R. et al. (2009) Targeting autophagy potentiates tyrosine kinase inhibitor-induced cell death in Philadelphia chromosome-positive cells, including primary CML stem cells. J. Clin. Invest. **119**, 1109–1123
43. Paglin, S., Hollister, T., Delohery, T., Hackett, N., McMahill, M., Sphicas, E., Domingo, D. and Yahalom, J. (2001) A novel response of cancer cells to radiation involves autophagy and formation of acidic vesicles. Cancer Res. **61**, 439–444
44. Amaravadi, R.K., Yu, D., Lum, J.J., Bui, T., Christophorou, M.A., Evan, G.I., Thomas-Tikhonenko, A. and Thompson, C.B. (2007) Autophagy inhibition enhances therapy-induced apoptosis in a Myc-induced model of lymphoma. J. Clin. Invest. **117**, 326–336
45. Zinn, R.L., Gardner, E.E., Dobromilskaya, I., Murphy, S., Marchionni, L., Hann, C.L. and Rudin, C.M. (2013) Combination treatment with ABT-737 and chloroquine in preclinical models of small cell lung cancer. Mol. Cancer **12**, 16
46. Saleem, A., Dvorzhinski, D., Santanam, U., Mathew, R., Bray, K., Stein, M., White, E. and DiPaola, R.S. (2012) Effect of dual inhibition of apoptosis and autophagy in prostate cancer. Prostate **72**, 1374–1381
47. White, E. and DiPaola, R.S. (2009) The double-edged sword of autophagy modulation in cancer. Clin. Cancer Res. **15**, 5308–5316
48. Tannenbaum, A. and Silverstone, H. (1949) The influence of the degree of caloric restriction on the formation of skin tumors and hepatomas in mice. Cancer Res. **9**, 724–727
49. Thompson, H.J., Zhu, Z. and Jiang, W. (2003) Dietary energy restriction in breast cancer prevention. J. Mammary Gland Biol. Neoplasia **8**, 133–142
50. Klurfeld, D.M., Welch, C.B., Davis, M.J. and Kritchevsky, D. (1989) Determination of degree of energy restriction necessary to reduce DMBA-induced mammary tumorigenesis in rats during the promotion phase. J. Nutr. **119**, 286–291
51. Colman, R.J., Anderson, R.M., Johnson, S.C., Kastman, E.K., Kosmatka, K.J., Beasley, T.M., Allison, D.B., Cruzen, C., Simmons, H.A., Kemnitz, J.W. and Weindruch, R. (2009) Caloric restriction delays disease onset and mortality in rhesus monkeys. Science **325**, 201–204
52. Rubinsztein, D.C., Marino, G. and Kroemer, G. (2011) Autophagy and aging. Cell **146**, 682–695
53. Dunn, S.E., Kari, F.W., French, J., Leininger, J.R., Travlos, G., Wilson, R. and Barrett, J.C. (1997) Dietary restriction reduces insulin-like growth factor I levels, which modulates apoptosis, cell proliferation, and tumor progression in p53-deficient mice. Cancer Res. **57**, 4667–4672
54. De Lorenzo, M.S., Baljinnyam, E., Vatner, D.E., Abarzua, P., Vatner, S.F. and Rabson, A.B. (2011) Caloric restriction reduces growth of mammary tumors and metastases. Carcinogenesis **32**, 1381–1387
55. Lashinger, L.M., Malone, L.M., Brown, G.W., Daniels, E.A., Goldberg, J.A., Otto, G., Fischer, S.M. and Hursting, S.D. (2011) Rapamycin partially mimics the anticancer effects of calorie restriction in a murine model of pancreatic cancer. Cancer Prev. Res. **4**, 1041–1051
56. Safdie, F., Brandhorst, S., Wei, M., Wang, W., Lee, C., Hwang, S., Conti, P.S., Chen, T.C. and Longo, V.D. (2012) Fasting enhances the response of glioma to chemo- and radiotherapy. PLoS ONE **7**, e44603
57. Raffaghello, L., Lee, C., Safdie, F.M., Wei, M., Madia, F., Bianchi, G. and Longo, V.D. (2008) Starvation-dependent differential stress resistance protects normal but not cancer cells against high-dose chemotherapy. Proc. Natl. Acad. Sci. U.S.A. **105**, 8215–8220
58. Helgason, G.V., Karvela, M. and Holyoake, T.L. (2011) Kill one bird with two stones: potential efficacy of BCR-ABL and autophagy inhibition in CML. Blood **118**, 2035–2043
59. Galavotti, S., Bartesaghi, S., Faccenda, D., Shaked-Rabi, M., Sanzone, S., McEvoy, A., Dinsdale, D., Condorelli, F., Brandner, S., Campanella, M. et al. (2013) The autophagy-associated factors DRAM1 and p62 regulate cell migration and invasion in glioblastoma stem cells. Oncogene **32**, 699–712

60. Michaud, M., Martins, I., Sukkurwala, A.Q., Adjemian, S., Ma, Y., Pellegatti, P., Shen, S., Kepp, O., Scoazec, M., Mignot, G. et al. (2011) Autophagy-dependent anticancer immune responses induced by chemotherapeutic agents in mice. Science **334**, 1573–1577
61. Maycotte, P., Aryal, S., Cummings, C.T., Thorburn, J., Morgan, M.J. and Thorburn, A. (2012) Chloroquine sensitizes breast cancer cells to chemotherapy independent of autophagy. Autophagy **8**, 200–212
62. McAfee, Q., Zhang, Z., Samanta, A., Levi, S.M., Ma, X.H., Piao, S., Lynch, J.P., Uehara, T., Sepulveda, A.R., Davis, L.E. et al. (2012) Autophagy inhibitor Lys05 has single-agent antitumor activity and reproduces the phenotype of a genetic autophagy deficiency. Proc. Natl. Acad. Sci. U.S.A. **109**, 8253–8258
63. Liu, J., Xia, H., Kim, M., Xu, L., Li, Y., Zhang, L., Cai, Y., Norberg, H.V., Zhang, T., Furuya, T. et al. (2011) Beclin1 controls the levels of p53 by regulating the deubiquitination activity of USP10 and USP13. Cell **147**, 223–234
64. Schafranek, L., Leclercq, T.M., White, D.L. and Hughes, T.P. (2013) Clarithromycin enhances dasatinib-induced cell death in chronic myeloid leukemia cells, by inhibition of late stage autophagy. Leuk. Lymphoma **54**, 198–201
65. Carella, A.M., Beltrami, G., Pica, G., Carella, A. and Catania, G. (2012) Clarithromycin potentiates tyrosine kinase inhibitor treatment in patients with resistant chronic myeloid leukemia. Leuk. Lymphoma **53**, 1409–1411

12

Autophagy as a defence against intracellular pathogens

Tom Wileman[1]

Department of Medicine, Norwich Medical School, University of East Anglia, Norwich NR4 7TJ U.K.

Abstract

Autophagy is a membrane trafficking pathway that results in the formation of autophagosomes which deliver portions of the cytosol to lysosomes for degradation. When autophagosomes engulf intracellular pathogens, the pathway is called 'xenophagy' because it leads to the removal of foreign material. Autophagy is activated during infection by Toll-like receptors that recognize pathogen-associated molecular patterns. This allows autophagy to kill micro-organisms and present pathogen components to the innate and acquired immune systems. The targeting of pathogens by autophagy is selective and involves a growing family of autophagy receptors that bind to the autophagosome membrane protein LC3 (light-chain 3)/Atg8 (autography-related protein 8). Ubiquitination of microbes identifies them as substrates for autophagy and they are delivered to autophagosomes by autophagy receptors that bind both ubiquitin and LC3/Atg8. Bacteria can also be detected before they enter the cytosol by autophagy receptors that scan the surface of membrane compartments for evidence of damage. The observation that some pathogens survive in cells suggests they can evade complete destruction by autophagy. For some bacteria this involves proteins that shield the surface of the bacteria from recognition by autophagy receptors. Other viruses and bacteria are resistant to degradation in lysosomes and use autophagosomes and/or lysosomes as sites for replication. Most of our current understanding of the role played by autophagy during microbial infection has come from studies of bacteria and viruses in tissue culture cell lines. Future work will focus on understanding how autophagy determines the outcome of infection '*in vivo*', and how autophagy pathways can be exploited therapeutically.

[1] email t.wileman@uea.ac.uk

Keywords:
galectin-8, innate immunity, NDP52, p62/sequestasome-like receptor, Toll-like receptor, virus, xenophagy.

Introduction

Early free-living eukaryotic cells needed to adapt to rapidly changing environments and uncertain food supply. Lack of nutrients imposed serious stress on these cells and this is thought to have driven the evolution of a membrane trafficking pathway called autophagy. Autophagy, which literally means 'self-eating' allows cells to deliver cytosolic organelles and proteins to lysosomes for degradation to provide a short-term supply of amino acids and this allowed early eukaryotes to generate the amino acids they needed to move and search for food. The capacity to degrade large quantities of cytoplasm also provided cells with a powerful mechanism to degrade intracellular pathogens. When autophagy engulfs pathogens, the pathway is called 'xenophagy' because it leads to the removal of foreign organisms [1]. Xenophagy therefore represents a very early stage in the evolution of innate immunity.

The process of autophagy

Several autophagy pathways have been described that deliver proteins to lysosomes. Microautophagy and chaperone-mediated autophagy deliver proteins directly from the cytoplasm into the lumen of the lysosome. Macroautophagy generates new membranes in cells called autophagosomes which engulf portions of the cytosol within double-membraned vesicles that fuse with lysosomes. Macroautophagy is important for the removal of pathogens and will be referred to as 'autophagy' in the rest of the chapter. At least 36 proteins are required for autophagy [2], and these are conserved from yeast through to fully differentiated mammalian cells. Autophagy is regulated by the TOR (target of rapamycin) kinase which senses amino acid levels in cells. When amino acids are abundant the TOR kinase inhibits autophagy and at the same time increases protein translation to increase cell mass. When food is scarce, amino acid levels fall and the TOR kinase is inactive. This slows translation and activates autophagy to generate amino acids through protein degradation. Autophagy is initiated by a class III PI3K (phosphoinositide 3-kinase) called vps34 that forms a complex with Beclin1/Atg6 to phosphorylate lipids at sites of autophagosome formation. Many membrane compartments can generate autophagosomes, but the best-characterized pathway involves the ER (endoplasmic reticulum) and/or mitochondria (Figure 1).

Anchoring of Atg14 to the ER and/or ER–mitochondrial contact sites [3–5], plays a key role in recruiting the vps34–Beclin1 complex from the cytosol to sites of autophagosome formation in response to starvation. Localized lipid phosphorylation generates small cup-shaped membranes called phagophores or isolation membranes which recruit effector proteins such as WIPI (WD-repeat protein interacting with phosphoinositides) 1/2 (Atg18) to prime phagophore expansion. The major structural protein of the autophagosome is called Atg8/LC3 and this is recruited from the cytosol to the phagophore following conjugation to PE (phosphatidylethanolamine). Generation of the LC3–PE conjugate (also known as LC3II) requires the Atg12–Atg5–Atg16 complex and results in phagophore expansion and eventual release of

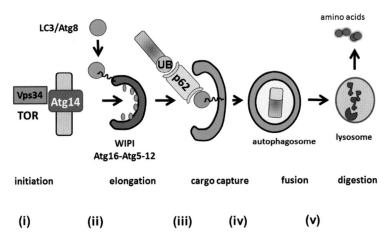

Figure 1. Generation of the autophagosomes
Autophagosome formation is regulated by the TOR kinase. When amino acid levels fall the TOR kinase is inhibited allowing Beclin 1 and the PI3K vps34 to initiate autophagosome formation on membranes enriched for Atg14 (**i**). Membrane expansion involves recruitment of WIPI, Atg5–Atg12–Atg16 complex and recruitment of Atg8/LC3 (**ii**). The membrane-bound form of LC3/Atg8 is called LC3II. WIPI and the Atg5–Atg12–Atg16 complex dissociate during autophagosome closure, but Atg8/LC3 remains with the autophagosome until fusion with lysosomes. Cargoes are captured by autophagy receptors, e.g. p62 that bind ubiquitin and LC3 (**iii**). Autophagosomes fuse with lysosomes to degrade proteins and organelles (**iv–v**). The amino acids enter the cytosol and activate the TOR kinase and this slows autophagy.

autophagosomes into the cytosol. During this stage, most of the proteins involved in autophagosome formation are recycled back into the cytosol. LC3II is the major protein of the autophagosome and remains with the autophagosome until fusion with the lysosomes. GFP tagging of LC3 (GFP–LC3) has become indispensable for visualizing autophagosomes in mammalian cells [6].

Microbes that evade lysosomes are captured by autophagy

Most bacteria enter cells by a combination of phagocytosis and endocytosis and are taken to lysosomes for degradation (Figure 2, i–ii). Lysosomes contain a wide range of hydrolytic enzymes maintained at acidic pH levels that are able to kill bacteria and increase presentation of microbial antigens to the immune system. The observation that several medically important pathogens survive in cells suggests they have evolved ways of evading direct delivery to lysosomes after endocytosis. Interestingly, recent work shows that these lysosome-evasion strategies trigger autophagy, allowing cells a second chance to deliver pathogens to lysosomes (reviewed in [7–10]). *Mycobacteria*, for example, remain in endocytic vesicles, but modify endosome/phagosome membranes so they cannot fuse with lysosomes (Figure 2, iii). These modified endosomes are taken up into autophagosomes and delivered to lysosomes. *Listeria*, *Shigella* and *Streptococci* secrete lysins to facilitate early release from endosomes and replicate in the cytosol (Figure 2, iv), but these cytosolic forms can be captured by autophagosomes. Some microbes, for example *Coxiella burnetii*, are resistant to degradation in lysosomes

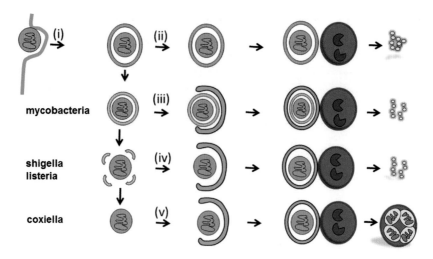

Figure 2. Autophagy pathways activated by intracellular microbes
Microbes enter the cell in endosomes (**i**) and for many, this results in delivery to lysosomes for degradation (**ii**). Microbes, for example mycobacteria, can evade delivery to lysosomes by modifying the endosome membrane (**iii**). These modified endosomes can be recognized by autophagosomes and delivered to lysosomes by xenophagy (**iii**). Other microbes, for example *Shigella*, secrete lysins to escape from the endosome into the cytosol (**iv**), the cytosolic bacteria can be recognized directly by autophagosomes and delivered to lysosomes. Some microbes, for example *Coxiella* (**v**), are resistant to degradation in lysosomes and activate xenophagy after delivery into the cytosol to gain access to the lysosome where they replicate and form parasitopherous vacuoles.

(Figure 2, v) and activate autophagy to gain access to the lysosome which they convert into a parasitophorus vacuole for replication. Intracellular microbes can also use parts of the autophagy pathway to promote replication through a process called 'non-canonical' autophagy [11]. For *Mycobacterium marinum* this involves generation of double-membraned vacuoles containing bacteria by pathways independently of Atg5 and LC3 [12]. *Brucella abortus* avoids delivery to lysosomes by generating a vacuole from the smooth ER. In common with autophagy, this requires localized lipid phosphorylation by PI3K, Beclin1/Atg6 and Atg14, but does not require later stages of autophagosome expansion powered by Atg5, Atg16L1 and LC3 [13]. FMDV (Foot and Mouth Disease virus) activates autophagy during cell entry by a non-canonical pathway which does not require PI3K [14], but is dependent on Atg5 and LC3.

Viruses infect cells by delivering genomes or nucleoprotein core particles into the cytoplasm, either directly through the plasma membrane, or following endocytosis. In contrast with studies of bacteria, relatively few studies have focused on the role played by autophagy in removing viruses immediately after they enter cells. Viruses are obligate intracellular pathogens and many viruses activate autophagy at the onset of viral genome replication. This may result from the recognition of viral genomes by cytosolic helicases such as RIG-I that bind viral RNA and trigger innate immunity to viral infection. Autophagy can also be activated by cell stress pathways, such as ER stress and UPRs (unfolded protein responses) that are activated when viruses use cellular membrane compartments as platforms for virus replication and synthesis of envelope proteins. Some viruses actually benefit from autophagy because the autophagosomes can become incorporated into sites of replication. This has been demonstrated for RNA viruses such as the poliovirus [15]. Autophagy can also inhibit virus replication and this can be important for controlling replication 'in vivo'. VSV (vesicular stomatitis virus) can infect insects and studies using fruitflies lacking genes essential for autophagy show that autophagy protects against infection [16]. SINV (Sindbis virus) is transmitted by mosquitoes and certain strains cause acute encephalitis in mice. Autophagy is activated during SINV infection leading to degradation of viral capsids. Studies using mice with neuron-specific loss of autophagy show a marked delay in the removal of viral antigens and increased susceptibility to SINV infection [17]. Similarly, neurovirulence of HSV1 (herpes simplex 1) is linked to expression of viral genes that bind Beclin1 and inhibit autophagy [18].

Autophagy is activated following recognition of pathogen-associated molecular patterns and damage signals

A striking feature emerging from recent studies is that recognition of intracellular pathogens by autophagy is linked to the exposure of DAMPs (damage-associated molecular patterns) and/or PAMPs (pathogen-associated molecular patterns). DAMPs are cell-derived molecules that generate 'danger' signals when they are displaced following cell stress or damage. Damage signals can be generated when cellular components are released from cells, or more subtly, when specific cellular compartments are ruptured during infection. PAMPs are pathogen-derived molecules that are only encountered during infection and examples include peptidoglycan and lipopolysaccharides generated by bacteria, the zymosan of yeast cell walls and dsRNA generated during viral infection. DAMPs and PAMPs are recognized by pattern recognition receptors such as the TLRs (Toll-like receptors), and this activates pro-inflammatory responses and/or production of interferon. Many of these signalling pathways also activate autophagy, allowing autophagy to play a crucial role in innate immunity (reviewed in [19]).

Many pathogens enter cells by endocytosis and/or phagocytosis allowing microbial PAMPs to be recognized by TLRs in endocytic compartments. Examples include detection of viral RNA in endosomes by TLR7 and recognition of bacterial lipopolysaccharide and lipopeptides by TLR1, TLR2 and TLR4. The precise mechanisms leading to activation of autophagy by TLRs remain to be understood, but one pathway appears to involve the TLR adaptor protein MyD88 (myeloid differentiation factor 88), which can activate Beclin1 by releasing it from an inhibitory complex with Bcl-2. The cytosol contains NOD (nucleotide-binding and oligomerization domain)-like receptors that bind PAMPs exposed by pathogens once they are released from endosomes. NOD1 and NOD2 proteins activate autophagy when they bind peptidoglycan within bacterial cell walls. NOD proteins bind autophagy protein Atg16L1 and may seed autophagosome formation at sites of bacterial entry resulting in efficient delivery of bacteria to lysosomes for degradation [20,21]. Genome-wide association studies have linked mutations in NOD2 and Atg16L1 to Crohn's disease making it possible that the inflammation associated with Crohn's disease is linked to defects in autophagy and microbial handling in intestinal epithelial cells. Recent studies show that mice that are deficient in autophagy in intestinal epithelial cells show greater sensitivity to *Salmonella* infection [22].

Selective autophagy involves autophagy receptors with LC3-interacting regions

Activation of autophagy in response to starvation is generally thought to lead to non-specific degradation of proteins and organelles. In contrast, activation of autophagy following recognition of pathogens leads to the selective degradation intracellular microbes. This has resulted in the discovery of a new class of innate immunity receptors called SLR (sequestasome-like receptor; or p62) or LIR (LC3-interacting region) proteins. The founding member of this family, p62/SQSTM1, was discovered during studies of inherited diseases where mutations lead to protein misfolding and aggregation. It had been known for many years that ubiquitin can target misfolded proteins for degradation by proteasomes, but many protein aggregates are too big for the proteasome and accumulate in the cytosol where they are degraded by autophagy. The targeting of aggregates for autophagy involves a family of linker proteins that can bind ubiquitin through UBDs (ubiquitin-binding domains), and at the same time bind LC3 in the autophagosome membrane using an LIR. Thus far, p62/SQSTM1 and NBR1 are the best studied and have clear roles in the removal of protein aggregates and damaged mitochondria [23].

Autophagy receptors control selective autophagy of intracellular pathogens

The role played by autophagy receptors in the control of intracellular pathogens is best illustrated through studies of *Salmonella typhimurium*. These bacteria enter cells in large endosomes which shrink to enclose one or two bacteria within membrane vesicles called a SCV (*Salmonella*-containing vacuole) (Figure 3). The bacteria then acquire autophagy markers LC3/Atg8, p62/SQSTM1 and ubiquitin suggesting uptake into autophagosomes. One signal for activation of autophagy is generated by the production of DAG (diacylglycerol) in response to secretion of bacterial proteins from the vacuole via the type 3 secretion system [24]. DAG acts as a lipid second messenger that promotes association of LC3/Atg8 with the SCV (Figure 3, i). A number of *Salmonella* escape from the vacuole, or reside in damaged vacuoles, exposing PAMPs to the cytosol, allowing access of autophagy receptors to the surface of the bacteria. Many pattern recognition proteins, for example TLRs and NOD-like receptors, contain LLR (leucine-rich) domains. This has stimulated a search for new LLR domain proteins that might play a role in linking pattern recognition with autophagy. A key goal has been to identify enzymes with E3 ubiquitin ligase activity that could transfer ubiquitin to the surface of bacteria. One such protein, LRSAM1, looks promising (Figure 3, ii). LRSAM1 is an LLR-domain protein with E3 ubiquitin ligase activity that binds an autophagy receptor called NDP52 (nuclear dot protein 52) that, in common with p62/SQSTM1, can bind ubiquitin and LC3 [25]. The LLR domain of LRSAM recognizes Gram-positive and Gram-negative bacteria and the enzyme can transfer ubiquitin to bacteria *in vitro*. A functional role for LRSAM has been demonstrated by knockdown experiments which show reduced autophagy of *Salmonella* in HeLa cells. Further evidence is provided from lymphocytes from patients suffering from Charcot–Marie–Tooth disease. These lymphocytes do not express LRSAM1 and are defective in ubiquitination of *Salmonella*.

Ubiquitin surrounding the bacteria may result in recognition by two (or more) autophagy receptors, p62/SQSTM1 and NDP52 (Figure 3, iii–iv) and both proteins contribute to elimination

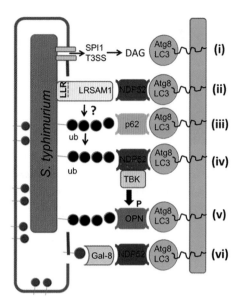

Figure 3. Recognition of *S. typhimurium* by autophagy receptors
Salmonella enter the cell by endocytosis and are retained in a modified endosome called the SCV. (**i**) Secretion of bacterial proteins into the cytosol through the type 3 secretion complex (SP1 T3SS) induces the formation of DAG which acts as a second messenger to activate autophagy. (**ii**) *Salmonella* rupture the SCV exposing the surface of the bacteria to the cytosol. LRSAM1 binds bacteria using an LRR domain and may use E3 ubiquitin ligase activity to transfer ubiquitin (ub, black spheres) to the surface of the *Salmonella*. LRSAM1 also binds autophagy receptor NDP52 which can bind LC3/Atg8. Ubiquitination can recruit autophagy receptors p62/SQSTM1 and NDP52 directly to the bacteria and link the microbe to LC3/Atg8 through LIR. (**iv**) NDP52 also binds TBK1 which forms an optineurin (OPN). Phosphorylation of optineurin by TBK allows optineurin to bind LC3/Atg8 (**v**). (**vi**) Damage to the SCV exposes sugars (red spheres) that would normally be hidden inside endosomes and lysosomes. These damage signals are recognized by galectin-8 (Gal-8) which binds NDP52.

of *Salmonella* [26]. NDP52 binds TBK [TANK (tumour-necrosis-factor-receptor-associated factor-associated nuclear factor-κB activator)-binding kinase] and optineurin (Figure 3, v). Optineurin also binds ubiquitin but has a low affinity for LC3; however, phosphorylation of optineurin by TBK generates a high-affinity-binding site for LC3, allowing TBK to play an important role in regulating autophagy of bacteria [26,27]. In addition to exposing PAMPs to the cytosol, disruption of the SCV by *Salmonella* exposes a newly discovered damage signal that is recognized by galectin-8 [28]. Galectin-8 is a cytosolic lectin that binds sugars that would normally be hidden on the inside of endosomes and lysosomes. When membrane compartments are damaged by pathogens the sugars provide an 'eat me' signal that recruits galectin-8 to the vacuole, the lectin then recruits autophagy receptor NDP52 to recruit LC3 (Figure 3, vi). This recent work on *S. typhimurium* has shown that recognition for autophagy is driven by pattern recognition and damage signals. This makes it likely that similar mechanisms will operate for any microbes that expose PAMPs to the cytosol or damage membrane compartments. This is illustrated by studies on *Shigella* and *Listeria* which secrete lysins to damage endosomes and are detected by galectin-8, NDP52 and p62/SQSTM1 [28,29].

Recognition of viruses for autophagy also involves autophagy receptors (Figure 4). Genetically distinct viruses such as SINV and FMDV, which are RNA viruses, and HSV1,

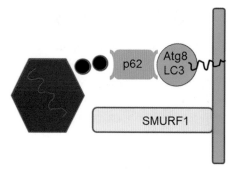

Figure 4. Recognition of viruses by autophagy receptors
Virus capsids in the cytosol may be recognized by p62/SQSTM1 and may therefore be ubiquitinated (black spheres). The autophagy receptor p62/SQSTM1 binds LC3/Atg8 allowing capture of viruses by autophagosomes. SMURF1 is also required for autophagy of SINV. SMURF1 is an E3 ubiquitin ligase that may recruit autophagosomes by binding directly to autophagosome lipids.

which is a DNA virus, are recognized by p62/SQSTM1 suggesting a common mechanism of detection [14,16,17]. Chikingunya virus capsids are toxic to cells and recognition of Chikingunya virus capsids by p62/SQSTM1 promotes cell survival by degrading capsids in autophagosomes [30]. A recent genome-wide silencing screen has identified SMURF1 (SMAD-specific E3 ubiquitin ligase 1) as a protein that is required for capture of SINV capsids by autophagosomes [31] and subsequent clearance of SINV and HSV1 from cells. SMURF1 has ubiquitin E3 ligase activity, but surprisingly, this is not required for selective autophagy, instead SMURF1 uses a membrane-targeting domain that may bind directly to autophagosome lipids.

Microbial evasion of autophagy

The observation that some pathogens survive in cells suggests they have found ways to avoid complete destruction by autophagy. *Listeria monocytogenese* and *Shigella flexneri* move rapidly within the cell using the actin cytoskeleton. *S. flexneri* secretes a protein called VirG to activate actin polymerization and VirG appears to activate autophagy because it binds the Atg5 protein involved in early stages of autophagosome formation. Virulent *Shigella* isolates may not activate autophagy because they generate the IcsB protein that blocks binding of VirG to Atg5 [32]. *L. monocytogenese* triggers autophagy when lysins rupture the endosome membrane to release the bacteria into the cytosol. Once in the cytosol, *L. monocytogenese* expresses the ActA protein to harness actin-based motility through recruitment of cellular Arp2/3 and Ena/VASP proteins. Early studies showed that autophagy markers are not recruited to mobile bacteria, suggesting that *L. monocytogenese* uses actin-based motility to escape from autophagy. Interestingly, strains that recruit Arp2/3 and Ena/VASP to polymerize actin, but do not move in the cell, also escape autophagosomes. This implies that polymerization of actin on the surface of the bacteria, rather than motility, can shield the surface of the microbe from recognition by autophagy receptors [33].

Conclusions and future research

It is becoming clear that autophagy plays a key role in innate immunity against infection. Selective autophagy of intracellular pathogens is triggered by detection of PAMPs and damage signals that recruit autophagy receptors to deliver pathogens, or vacuoles containing them, to lysosomes for degradation. Delivery to lysosomes kills microbes and increases presentation of microbial antigens to the immune system. Recent experiments suggest that the recruitment of autophagy receptors may vary between pathogens [28]. It will be important to determine if this reflects targeting of microbes to different autophagy pathways, for example, through binding to specific ATG8/LC3 family members [34], or the fine tuning of microbial-autophagosome evasion strategies. Ubiquitin plays a key role in recruiting autophagy receptors during selective autophagy. Studies on cargo capture need to be extended to identify the E3 ubiquitin ligases that mark microbes for ubiquitination and the microbial proteins that are recognized for ubiquitination. Galectin-8 and DAG provide examples of non-ubiquitin-based signals for targeting bacteria for autophagy, there may be other signals on the basis of recognition of membrane damage. Studies in cell culture need to be extended to animal models to determine the effects of autophagy on the outcome of infection. Conditional expression of autophagy proteins and/or autophagy receptor proteins can be used to determine the role played by autophagy in immunological surveillance and determining virulence and pathogenesis. Recent genome-wide screens have linked mutations in autophagy proteins to human diseases. Crohn's disease is linked to mutations in NOD sensors and autophagy protein Atg16L, and mutations in autophagy receptor p62/SQSTM1 are linked to Paget's disease. It will be important to determine whether these mutations cause disease because they compromise microbial handling by autophagy.

Summary

- Autophagy evolved as a response to starvation and results in the formation of autophagosomes that engulf portions of the cytosol and fuse with lysosomes.
- When autophagy engulfs intracellular microbes the pathway is called xenophagy, indicating the removal of foreign material.
- Autophagy is emerging as an important arm of anti-microbial immunity because lysosomes are able to kill microbes and increase presentation of microbial antigens to the innate and acquired immune systems.
- Activation of autophagy during infection and recognition of intracellular pathogens is linked to the exposure of DAMPs and/or PAMPs.
- Recognition of pathogens is selective and involves a family of autophagy receptors that bind the major membrane protein of the autophagosome LC3/Atg8.
- Autophagy receptors bind microbes that have been tagged with ubiquitin, whereas others scan the surface of membrane compartments containing microbes for evidence of damage.

- Many pathogens have evolved autophagy-evasion strategies that shield them from recognition by autophagy receptors. Viruses and bacteria that are resistant to degradation in lysosomes can use autophagosomes and/or lysosomes as sites for replication.

References

1. Levine, B. (2005) Eating oneself and uninvited guests: autophagy-related pathways of cellular defence. Cell **120**, 159–162
2. Motley, A.M., Nuttall, J.M. and Hettema, E.H. (2012) Pex3-anchored Atg36 tags peroxisomes for degradation in *Saccharomyces cerevisiae*. EMBO J. **31**, 2852–2868
3. Hamasaki, M., Furuta, N., Nezu, A., Yamamoto, A., Fujita, N., Oomori, H., Noda, T., Haraguchi, T., Hiraoka, Y., Amano, A. and Yoshimori, T. (2013) Autophagosomes form at ER–mitochondria contact sites. Nature **495**, 389–393
4. Axe, E.L., Walker, S.A., Manifava, M., Chandra, P., Roderick, H.L., Habermann, A., Griffiths, G. and Ktistakis, N.T. (2008) Autophagosome formation from membrane compartments enriched in phosphatidylinositol 3-phosphate and dynamically connected to the endoplasmic reticulum. J. Cell Biol. **182**, 685–701
5. Hailey, D.W., Rambold, A.S., Satpute-Krishnan, P., Mitra, K., Sougrat, R., Kim, P.K. and Lippincott-Schwartz, J. (2010) Mitochondria supply membranes for autophagosome biogenesis during starvation. Cell **141**, 656–667
6. Klionsky, D., Abdalla, F.C., Abeliovich, H., Abraham, R.T., Acevedo-Arozena, A., Adeli, K., Agholme, L., Agnello, M., Agostinis, P., Aguirre-Ghiso, J.A. et al. (2012) Guidelines for the use and interpretation of assays for monitoring autophagy. Autophagy **8**, 445–544
7. Mostowy, S. and Cossart, P. (2012) Bacterial autophagy: restriction or promotion of bacterial replication? Trends Cell Biol. **22**, 283–291
8. Cemma, M. and Brumell, J.H. (2012) Interactions of pathogenic bacteria with autophagy systems. Curr. Biol. **22**, R540–R545
9. Pierini, R., Cottam, E., Roberts, R. and Wileman, T. (2009) Modulation of membrane traffic between endoplasmic reticulum, ERGIC and Golgi to generate compartments for the replication of bacteria and viruses. Semin. Cell Dev. Biol. **7**, 828–833
10. Randow, F. and Munz, C. (2012) Autophagy in the regulation of pathogen replication and adaptive immunity. Trends Immunol. **33**, 475–487
11. Codogno, P., Mehrpour, M. and Proikas-Cezanne, T. (2012) Canonical and non-canonical autophagy: variations on a common theme of self-eating. Nat. Rev. Mol. Cell Biol. **13**, 7–12
12. Collins, C.A., Maziere, A.D., van Dijk, S., Carlsson, F., Klumperman, J. and Brown, E.J. (2009) Atg5-independent sequestration of ubiquitinated mycobacteria. PLoS Pathogen **5**, e1000430
13. Starr, T., Child, R., Wehrly, T.D., Hansen, B., Hwang, S., Lopez-Otin, C., Virgin, H.W. and Celi, J. (2012) Selective subversion of autophagy complexes facilitates completion of the *Brucella* intracellular cycle. Cell Host Microbe **11**, 33–45
14. Berryman, S., Brooks, E., Burman, A., Hawes, P., Roberts, R., Netherton, C., Monaghan, P., Whelband, M., Cottam, E., Elazar, Z. et al. (2012) Foot-and-mouth disease virus induces autophagosomes during cell entry via a class III phosphatidylinositol 3-kinase-independent pathway. J. Virol. **86**, 12940–12953
15. Jackson, W.T., Giddings, T.H., Taylor, M.P., Mulinyawe, S., Rabinovitch, M., Kopito, R.R. and Kirkegaard, K. (2005) Subversion of cellular autophagosomal machinery by RNA viruses. PLoS Biol. **3**, e156
16. Shelly, S., Lukinova, N., Berman, S. and Cherry, A. (2009) Autophagy is an essential component of *Drosophila* immunity against vesicular stomatitis virus. Immunity **30**, 588

17. Orvedahl, A., MacPherson, S., Sumpter, R.J., Talloczy, Z. and Levine, B. (2010) Autophagy protects Sindbis virus infection of the central nervous system. Cell Host Microbe **7**, 115–127
18. Orvedahl, A., Alexander, D., Talloczy, Z., Sun, Q., Wei, Y., Zhang, W., Burns, D., Leib, D.A. and Levine, B. (2007) HSV-1 ICP34.5 confers neurovirulence by targeting the Beclin 1 autophagy protein. Cell Host Microbe **1**, 23–35
19. Deretic, V. (2011) Autophagy in immunity and cell autonomous defense against microbes. Immunological Rev. **240**, 92–104
20. Travassos, L.H., Carneiro, L.A.M., Ramjeet, M., Hussey, S., Kim, Y.-G., Magalhas, J.G., Yuan, L., Soares, F., Chea, E., Le Bourhis, L. et al. (2009) Nod1 and Nod2 direct autophagy by recruiting ATG16L1 to the plasma membrane at the site of bacterial entry. Nat. Immunol. **11**, 55
21. Cooney, R., Baker, J., Brain, O., Danis, B., Pichulik, T., Ferguson, D.J.P., Campbell, B.J., Jewell, D. and Simmons, A. (2010) NOD2 stimulation induces autophagy in dendritic cells influencing bacterial handling and antigen presentation. Nat. Med. **16**, 90
22. Benjamin, J.L., Sumpter, R., Levine, B. and Hooper, L.V. (2013) Intestinal epithelial autophagy is essential for host defense against invasive bacteria. Cell Host Microbe **13**, 828–833
23. Kirkin, V., McEwan, D.G., Novak, I. and Dikic, I. (2009) A role for ubiqutin in selective autophagy. Mol. Cell **34**, 259
24. Shahnazari, S., Yen W-L, Birmingham, C.L., Shiu, J., Namolovan, A., Zheng, Y.T., Nakayama, K., Klionsky, D. and Brumell, J.H. (2010) A diacylglycerol-dependent signalling pathway contributes to regulation of antibacterial autophagy Cell Host Microbe **8**, 137–146
25. Huett, A., Heath, R.J., Begun, J., Sassi, S.O., Baxt, L.A., Vyas, J.M., Goldberg, M.B. and Xavier, R.J. (2012) The LRR and RING domain protein LRSAM1 is an E3 ligase crucial for ubiquitin-dependent autophagy of intracellular *Salmonella typhimurium.* Cell Host Microbe **12**, 778–790
26. Thurston, T.L.M., Ryzhakov, G., Bloor, S., Muhlinen, N. and Randow, F. (2009) The TBK1 adaptor and autophagy receptor NDP52 restricts the proliferation of ubiquitin-coated bacteria. Nat. Immunol. **10**, 1215–1221
27. Wild, P., Farham, H., McEwan, D.G., Wagner, S., Rogov, V.V., Brady, N.R., Richter, B., Korac, J., Waidmann, O., Choudhary, C. et al. (2011) Phosphorylation of the autophagy receptor optineurin restricts *Salmonella* growth. Science **333**, 228–233
28. Thurston, T.L.M., Wandel, M.P., Muhlinen, N., Foeglein, A. and Randow, F. (2012) Galectin-8 targets damaged vesicles for autophagy to defend cells against bacterial invasion. Nature **482**, 414–419
29. Mostowy, S., Sancho-Shimizu, V., Hamon, M.A., Simeone, R., Brosch, R., Johansen, T. and Cossart, P. (2011) P62 and NDP52 proteins target intracytosolic *Shigella* and *Listeria* to different autophagy pathways. J. Biol. Chem. **286**, 26987–26995
30. Judith, D., Mostowy, S., Bourai, M., Gangneux, N., Lelek, M., Lucas-Hourani, M., Cayet, N., Jacob, Y., Prévost, M.-C., Pierre, P. et al. (2013) Species-specific impact of the autophagy machinery on Chikungunya virus infection. EMBO Rep. **14**, 534–544
31. Orvedahl, A., Sumpter, R., Xiao, G., Ng, A., Zou, Z., Tang, Y., Narimatsu, M., Gilpin, C., Sun, Q., Roth, M. et al. (2011) Image-based genome wide siRNA screen identifies selective autophagy factors. Nature **480**, 113–117
32. Ogawa, M., Yoshimori, T., Suzuki, T., Sagara, H., Mizushima, N. and Sasakawa, C. (2005) Escape of intracellular shigella from autophagy. Science **307**, 727–731
33. Yoshikawa, Y., Ogawa, M., Hain, T., Yoshida, M., Fukumatsu, M., Kim, M., Mimuro, H., Nakawaga, I., Yanagawa, T., Ishii, T. et al. (2009) Listeria monocytogenes ActA-mediated escape from autophagic recognition. Nat. Cell Biol. **11**, 1233–1240
34. von Muhlinen, N., Akutsu, M., Ravenhill B J. Foeglein, A., Bloor, S., Ruthgerford, T.J., Freund, S.M.V., Komander, D. and Randow, F. (2012) LC3C, bound selectively by a non-canonical LIR motif in NDP52, is required for antibacterial autophagy. Mol. Cell **48**, 329–342

INDEX

A

actin, 59, 65, 67, 69, 70, 71, 75, 88, 160
ageing, 119–129
aggrephagy, 57, 79, 81, 83, 85, 86, 87, 88
amino-acid starvation, 1, 3, 4, 5, 6, 7, 8, 12, 31, 34, 35
γ-aminobutyric acid receptor-associated protein (see GABARAP)
AMP-activated protein kinase (see AMPK)
AMPK, 1, 6, 8, 9, 10, 18, 32, 120, 121, 122, 123, 124, 125, 126, 134, 135, 137, 138, 144
anticancer therapy, 133, 139, 145, 146
ATG protein, 2, 3, 32, 33, 35, 39, 113
ATG1, 1, 3, 4, 5, 6, 7, 8, 9, 10, 11, 12, 18, 31, 40, 57, 135
ATG3, 33, 39, 41, 42, 44, 45, 47, 53, 54, 55, 62, 113, 134, 135, 136
ATG4, 34, 39, 41, 46, 47, 53, 54, 55, 56, 60, 61, 62, 134, 136
ATG5, 2, 3, 19, 20, 32, 33, 34, 39, 40, 41, 42, 44, 45, 46, 47, 48, 54, 55, 62, 72, 74, 107, 112, 113, 120, 121, 122, 134, 136, 139, 154, 155, 156, 160
ATG8, 2, 3, 6, 7, 19, 33, 39, 40, 41, 42, 43, 44, 46, 47, 51–62, 67, 69, 79, 80, 82, 83, 84, 85, 88, 89, 94, 95, 96, 120, 135, 153, 154, 155, 158, 159, 160, 161
ATG9, 8, 10, 11, 19, 30, 33, 34, 40
ATG10, 39, 41, 42, 47, 54, 134, 135, 136
ATG12, 2, 3, 19, 32, 33, 34, 39, 40, 41, 42, 44, 45, 46, 47, 54, 55, 62, 72, 74, 112, 113, 120, 121, 134, 135, 136, 139, 147, 154, 155
ATG13, 1, 5, 6, 18, 19, 31, 134, 135
ATG14, 19, 22, 23, 32, 35, 40, 46, 154, 155, 156
ATG16, 19, 32, 33, 34, 39, 40, 42, 45, 46, 47, 54, 55, 62, 69, 73, 134, 136, 154, 155
ATG17, 1, 5, 6, 7, 31, 135

ATG32, 81, 82, 93, 94, 95, 98, 101
ATG101, 1, 2, 3, 5, 7, 18, 19, 134, 135
autolysosome, 29, 30, 43, 46, 67, 68, 106, 134, 136
autophagic cell death, 105, 107, 108
autophagosome, 2, 3, 4, 6, 7, 8, 11, 12, 18, 19, 20, 21, 22, 23, 24, 25, 29, 30, 31, 32, 33, 34, 35, 36, 39, 40, 42, 43, 44, 46, 47, 51, 52, 53, 54, 55, 56, 57, 59, 60, 61, 62, 65, 66, 67, 68, 69, 70, 71, 72, 73, 74, 75, 79, 80, 83, 89, 94, 95, 97, 98, 99, 106, 108, 109, 112, 113, 119, 120, 121, 126, 128, 133, 134, 135, 136, 147, 153, 154, 155, 156, 158, 160, 161
autophagosome formation, 3, 5, 6, 7, 11, 12, 18, 19, 20, 22, 23, 24, 25, 31, 34, 39, 40, 43, 44, 46, 47, 51, 53, 56, 57, 59, 61, 67, 68, 73, 74, 75, 106, 108, 109, 112, 126, 128, 135, 154, 155, 157, 160
autophagy-mediated protection, 105

B

BAD, 112, 137, 138, 139
BAK, 107, 108, 112, 138
BAX, 107, 108, 112, 113, 138
Beclin1, 2, 3, 7, 8, 11, 17, 19, 20, 22, 23, 32, 33, 34, 35, 107, 112, 113, 120, 122, 126, 127, 128, 134, 135, 137, 139, 147, 154, 156, 157
Bcl-2, 22, 23, 110, 112, 113, 134, 135, 137, 138–139, 157
Bcl-2/Bcl-xL-antagonist, causing cell death (see BAD)
Bcl-2 homology 3 interacting-domain death agonist (see BID)
Bcl-2 interacting killer (see BIK)
Bcl-2-interacting mediator of cell death (see BIM)
Bcl-xL, 112, 135, 137, 138
BID, 110, 138

BIK, 137, 138
BIM, 113, 137, 138, 139
BNIP3, 81, 82, 95, 112, 121, 137, 138

C

caspase, 105, 108, 109, 110, 111, 114
chaperone-mediated autophagy
   (see CMA)
chloroquine, 109, 111, 112, 133, 139, 142, 143, 145, 146
chronic myeloid leukaemia, 133, 139, 141, 145
conjugation, 3, 19, 32, 33, 39, 40, 41, 42, 43, 44, 46, 47, 54, 56, 62, 85, 134, 135, 136, 147, 154
CMA, 2, 18, 31, 106, 120,

D

dBRUCE, 105, 108, 109, 114
deconjugation, 39, 41, 46, 47, 57, 61, 62
DFCP1, 2, 3, 4, 19, 20, 21, 22, 32, 34, 35
double FYVE domains-containing protein 1 (see DFCP1)
Drosophila baculovirus inhibitor of apoptosis repeat-containing ubiquitin-conjugating enzyme (see dBRUCE)

E

endosome, 65, 67, 72, 73, 74, 75, 79, 155, 156, 159, 160
erythropoiesis, 93, 95

F

FIP200, 1, 2, 3, 5, 7, 10, 18, 19, 31, 33, 134, 135, 139
focal adhesion kinase family-interacting protein 200 kDa (see FIP200)
FUN14 domain containing 1 (see FUNDC1)
FUNDC1, 80, 82, 98
fusion, 3, 24, 34, 35, 42, 43, 44, 46, 53, 54, 56, 57, 59, 60, 61, 65, 67, 68, 69, 70, 71, 72, 73, 74, 75, 76, 87, 98, 99, 100, 101, 106, 113, 120, 134, 136, 145, 147, 155

G

GABARAP, 33, 39, 40, 42, 51, 52, 53, 54, 56, 62, 80, 84, 95
galectin-8, 88, 154, 159, 161

GATE-16, 33, 51, 53, 54, 56, 62
Golgi-associated ATPase enhancer of 16 kDa (see GATE-16)

H

haemopoietic stem cell, 133, 139, 145
hydroxychloroquine, 133, 140, 141, 142, 143, 145, 146, 147

I

innate immunity, 154, 156, 157, 158, 161

L

LC3, 7, 19, 20, 22, 33, 34, 39, 40, 42, 46, 51, 52, 53, 54, 55, 56, 57, 59, 60, 62, 67, 68, 69, 70, 71, 79, 80, 81, 83, 84, 85, 88, 89, 95, 96, 97, 98, 100, 112, 120, 121, 122, 134, 135, 136, 147, 153, 154, 155, 156, 158, 159, 160, 161
lysosome, 2, 17, 18, 20, 22, 23, 24, 25, 30, 31, 40, 51, 52, 54, 59, 60, 65, 66, 67, 68, 69, 70, 71, 72, 73, 74, 75, 79, 87, 88, 93, 94, 96, 105, 106, 107, 109, 111, 112, 113, 119, 120, 121, 126, 127, 128, 133, 134, 136, 145, 147, 153, 154, 155, 156, 157, 159, 161, 162

M

MAM, 30, 35
mammalian/mechanistic target of rapamycin (see mTOR)
mammalian/mechanistic target of rapamycin complex 1 (see mTORC1)
maturation, 3, 6, 22, 23, 24, 53, 56, 57, 61, 65–76, 88, 95, 121, 134, 135, 147
MCL-1, 113, 137, 138
microtubule, 7, 11, 19, 53, 65, 69, 70, 71, 87, 120, 121, 125, 128, 135
microtubule-associated protein 1 light-chain 3 (see LC3)
mitochondria-associated endoplasmic reticulum membrane (see MAM)
mitochondrial
   dynamics, 93, 98, 99, 100
   dysfunction, 93, 96, 101, 124, 125, 128

© 2013 Biochemical Society

mTOR, 7, 9, 18, 95, 120, 121, 125, 126, 136, 138, 140, 144
mTORC1, 1, 5, 6, 7, 8, 9, 10, 18, 19, 31, 32, 88, 120, 125, 126, 129, 134, 135, 136, 137, 138, 144, 147
myeloid cell leukaemia sequence 1 (see MCL-1)
myotubularin, 24

N

NBR1, 79, 80, 81, 83, 87, 88, 158
NDP52, 57, 79, 80, 81, 83, 84, 85, 88, 154, 158, 159
neighbour of BRCA1 (breast cancer early-onset 1) gene 1 (see NBR1)
neurodegeneration, 52, 93, 101, 120, 126–128
Nix, 57, 93, 95, 96, 98, 101
Noxa, 137, 138
nuclear dot protein 52 (see NDP52)

O

omegasome, 17–25, 30, 34, 35
optineurin, 79, 81, 83, 88, 159

P

p53, 10, 108, 112, 121, 123, 126, 135, 136, 137, 138, 140, 144
p53 up-regulated modulator of apoptosis (see PUMA)
p62/sequestosome-like receptor, 57, 80, 81, 83, 158, 159, 160, 161
parkin, 93, 96, 97, 98, 99, 100, 101, 127, 128
Parkinson's disease, 86, 93, 96, 126, 127, 128, 129
PAS, 1, 5, 6, 7, 18, 19, 30, 31, 33, 34, 36, 40, 42, 43, 45, 46
phagophore, 2, 3, 7, 18, 22, 24, 30, 31, 32, 33, 34, 35, 36, 40, 51, 52, 56, 68, 69, 75, 80, 84, 85, 134, 135, 154
phosphatidylethanolamine, 2, 3, 19, 20, 21, 33, 39, 40, 41, 42, 43, 44, 45, 46, 47, 52, 53, 54, 55, 56, 57, 59, 62, 134, 136, 154
phosphatidylinositol 3-phosphate (see PI3P)

PI3P, 2, 3, 7, 17, 19, 20, 21, 22, 23, 24, 25, 32, 46, 67, 71, 73, 74
PINK1, 93, 96, 97, 100, 101, 128
pre-autophagosome structure (see PAS)
protein aggregation, 80, 86, 119, 120, 127, 128
PTEN (phosphatase and tensin homologue deleted on chromosome 10)-induced putative kinase 1 (see PINK1)
PUMA, 137, 138

R

Rab, 23, 57, 65, 67, 70, 71, 72–73, 74, 75, 76, 120, 121
reactive oxygen species (see ROS)
ROS, 61, 86, 94, 95, 107, 120, 124, 125, 126, 139
Rubicon, 22, 23, 67, 73, 134, 135

S

selective autophagy, 31, 40, 71, 79–89, 94, 158, 160, 161
senescence, 120, 125, 126
SIRT1, 121, 122, 123, 124, 126, 144
SNARE, 34, 35, 44, 53, 61, 65, 67, 71, 72, 74–75, 76, 120, 121

T

TANK (tumour-necrosis-factor-receptor-associated factor-associated nuclear factor-κB activator)-binding kinase (see TBK1)
TBK1, 79, 88, 159
tether, 43, 65, 67, 72, 73, 74
Toll-like receptor, 154, 157, 158
TRAIL, 105, 110, 111, 114
tumorigenesis, 133, 136, 137, 139
tumour-necrosis-factor-related apoptosis-inducing ligand (see TRAIL)

U

ubiquitin, 32, 33, 39–47, 51, 52, 53, 54, 55, 57, 63, 67, 69, 79, 80, 81, 82, 83, 84, 85, 86, 87, 88, 89, 96, 97, 98, 99, 105, 106, 108, 122, 128, 134, 135, 153, 155, 158, 159, 160, 161
ubiquitin-like protein, 32, 39, 41, 44, 45, 51, 53, 69, 135

ULK, 1, 3, 5, 6, 7, 8, 9, 10, 11, 17, 18, 19, 25, 31, 33, 34, 35, 57, 69, 88, 120, 121, 134, 135
ULK complex, 1, 2, 5, 7, 8, 10, 11, 17, 18, 19, 33, 34, 134, 135, 147
uncoordinated-51-like kinase (see ULK)
UVRAG, 22, 23, 67, 73, 74, 112, 113, 134, 135, 136
UV radiation resistance-associated gene (see UVRAG)

V

vacuole, 6, 40, 42, 46, 51, 52, 57, 59, 60, 70, 71, 79, 80, 83, 88, 94, 95, 159, 158, 159, 160

virus, 23, 47, 57, 81, 153, 154, 156, 157, 159, 160
Vps34, 17, 18, 19, 20, 21, 22, 23, 25, 32

W

WD-repeat protein interacting with phosphoinositides (see WIPI1/2)
WIPI1/2, 2, 3, 19, 22, 32, 33, 34, 40, 154, 155

X

xenophagy, 31, 57, 79, 81, 83, 85, 88, 153, 154, 156, 161